An Introduction to HPLC for

Pharmaceutical Analysis

MTS

MOURNE TRAINING SERVICES

Mourne Training Services
14 Burren Road, Warrenpoint, Co. Down, Northern Ireland, BT34 3SA, United Kingdom
Tel: +44 (0)28 30268236
Email: info@mournetrainingservices.co.uk
Web: www.mournetrainingservices.co.uk

ISBN   978-0-9561528-0-0

Preface

This book is aimed at the analyst who is new to HPLC. The content is written specifically for HPLC applied to pharmaceutical analysis. The purpose of the book is to provide the necessary information which will enable the reader to confidently follow a HPLC analytical method. Therefore, the content is based around the information that an analyst needs to be able to use HPLC in a pharmaceutical analysis environment. The concepts of how the technique works and how it is applied in practice are introduced in a logical order, the intention being to build up an understanding of the whole technique gradually.

After an introduction to the topics of both HPLC and pharmaceuticals in chapter 1, the next three chapters deal with the equipment and instrumentation required for the technique. Chapter 2 describes the stationary phase and enables the reader to make sense of the many parameters used to describe a HPLC column. Chapter 3 provides all the necessary information to enable correct preparation and use of HPLC mobile phase. Chapter 4 introduces the equipment which is used to implement the technique of HPLC. The key features of each of the separate parts of the system are described to provide an understanding of how they combine to perform analysis. The second half of the book, chapters 5 to 8, concentrates on the application of HPLC for pharmaceutical analysis. Chapter 5 describes the analytical method and how to interpret its contents. Chapter 6 is a step-by-step guide on how to follow a method and set up a HPLC analysis. Chapter 7 explains the meaning of system suitability criteria and how to interpret the values obtained during an analysis. The final chapter explains the common methods of calibration and quantification used for pharmaceutical analysis, this will enable the reader to calculate the results from a HPLC analysis correctly.

In addition to providing an introduction to HPLC for pharmaceutical analysis it is intended that this book will be a useful resource. At the end of each chapter there is a list of references and/or further reading which will help the reader to develop their expertise in the technique. Useful data is provided throughout the book, such as: buffers and their pKas; conversion tables for units of pressure; and lists of the UV cutoffs for common solvents and buffers. There is a glossary and a list of abbreviations at the back of the book to help the reader become familiar with the terminology used in HPLC and pharmaceutical analysis. When a new term is introduced it is shown in **bold** to indicate to the reader that a definition is available in the glossary.

Due to the applied nature of this topic it is necessary to mention the names of chromatography products, suppliers and manufacturers throughout the book. No endorsement of these products is intended. Every effort has been made to ensure that brands and trademarks are accurately assigned.

I hope that this book provides the reader with the information they need to get started in HPLC. When writing the text and deciding what to include, my approach was to include the information that I wished someone had told me at the beginning of my career in HPLC. I have also included recent developments in HPLC technology to ensure that the content is fully up to date. There is a lot of information to take in at a first reading in these eight chapters, my wish is that this book will be a useful reference for the reader as they gain experience in HPLC analysis.

Oona McPolin

Contents

CHAPTER 5

THE HPLC ANALYTICAL METHOD 65

CHAPTER 6

PERFORMING HPLC ANALYSIS 83

Introduction

The analytical technique of High Performance Liquid Chromatography (HPLC) is used extensively throughout the pharmaceutical industry. It is used to provide information on the composition of drug related samples. The information obtained may be **qualitative**, indicating what compounds are present in the sample or **quantitative**, providing the actual amounts of compounds in the sample. HPLC is used at all the different stages in the creation of a new drug, and also is used routinely during drug manufacture. The aim of the analysis will depend on both the nature of the sample and the stage of development. HPLC is a chromatographic technique, therefore it is necessary to have a basic understanding of chromatography to understand how it works.

What is Chromatography?

A Russian botanist, Mikhail Tswett (1872 - 1919), is credited with the first use of chromatography in 1906 when he separated plant pigments such as chlorophylls and xanthophylls. He passed them through a glass column packed with calcium carbonate. These pigments are coloured and thus the technique was named using the Greek terms, 'chroma' meaning 'colour', and 'graphein' meaning 'to write'. This explains why the name seemingly bears little relation to the use of the technique today.

Chromatography is a technique which separates components in a mixture due to the differing time taken for each component to travel through a stationary phase when carried through it by a mobile phase. The possible mixtures of phases give rise to the types of chromatography listed in Table 1.

Table 1   Types of chromatography

Type of chromatography	Mobile Phase	Stationary Phase
Gas Chromatography	Gas	Solid/Liquid
Liquid Chromatography	Liquid	Solid/Liquid
Supercritical-fluid chromatography	Supercritical fluid	Solid/Liquid

The stationary phase is fixed in place either in a column (a hollow tube made out of a suitable material, e.g. glass) or on a planar surface and the mobile phase moves over or through the stationary phase carrying with it the sample of interest. In practice the stationary phase can be a solid, a liquid adsorbed on a solid or an organic species (e.g. a C_{18} alkyl chain) bonded to a solid surface. In gas chromatography and supercritical-fluid chromatography the stationary phase is fixed in place in a column. In liquid chromatography the stationary phase may be fixed in place either in a column or on a planar surface. In HPLC a column is used. The name given to liquid chromatography on a planar surface is Thin Layer Chromatography (TLC).

Figure 1 shows a very simplistic representation of how the separation is achieved. A mixture of component A and component B is introduced to the mobile phase. A and B are travelling at the same rate as the rate of flow of the mobile phase. At time t1 they encounter the stationary phase. Both A and B are attracted to the stationary phase and this slows down their rate of travel in relation to the rate of the mobile phase. This occurs because both A and B are in an equilibrium between time spent on the stationary phase and time spent in the mobile phase. Time spent on the stationary phase does not result in travel of the component through the stationary phase, only time spent in the mobile phase allows travel.

A has a slightly greater affinity for the stationary phase than B. This means that relative to B, A spends more time on the stationary phase and travels at a slower rate than B. At time t2 A and B are beginning to separate. At time t3 they are fully separated. The extra time taken for B to reach the end of the stationary phase at t4 results in further separation of A and B.

Figure 1   Representation of the separation of two components, A and B

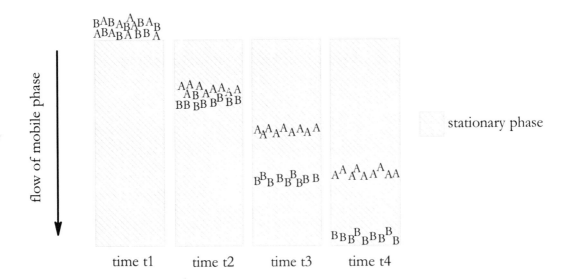

The time taken for a component to travel through the stationary phase is referred to as the **retention time**, thus the retention time of B is equal to time t4. The process of

the sample being carried continuously in the mobile phase is known as **elution**, B elutes at time t4. Because A and B migrate at different rates through the stationary phase, the term **differential migration** is used to describe the separation process.

High Performance Liquid Chromatography (HPLC)

In liquid chromatography the mobile phase described previously is always in the liquid phase. High Performance Liquid Chromatography (HPLC) is characterised by the use of very finely divided stationary phase (very small particles) through which mobile phase is pumped at high pressure. The name 'high performance' liquid chromatography originates from the late 1960s, when technology was developed to make both stationary phase column **packings** of very small particle size and sophisticated instruments which could generate the high pressures required for the flow of mobile phase through these packings. 'High performance' was used to distinguish the new technique from previous liquid chromatographic procedures involving glass columns packed with larger particle size material. These previous techniques were slower and did not have the reproducibility required for analytical purposes, however they are still much in use for preparative purposes. The name 'high *pressure* liquid chromatography' has also been used to describe the technique due to the high pressures involved but the term 'performance' is usually preferred. The development of the column packings and instrumentation for HPLC has progressed to such an extent that it is now relatively easy to perform analyses.

The Types of HPLC

There are a number of different types of HPLC, these include **partition, adsorption, ion-exchange** and **size-exclusion** chromatography. These different types tend to be complementary and most drug-related samples may be analysed using one or more of them.

Partition Chromatography

In partition chromatography the stationary phase is an organic species covalently bonded to the packing. During analysis the equilibrium experienced by the component of interest is a partition between the liquid mobile phase and the bonded stationary phase. There are hundreds of different **bonded phases** available commercially. These have a wide range of chemistries to enable the separation and analysis of a wide variety of mixtures. Examples of bonded phases include alkyl groups such as C_{18} and C_8. Partition is the most widely used type of HPLC in pharmaceutical analysis. The majority of bonded phases are **non-polar** and they are typically combined with a **polar** mobile phase.

Adsorption Chromatography

In adsorption chromatography the stationary phase is the surface of a finely divided solid such as silica or alumina. During analysis the analyte and the mobile phase compete for sites on the surface of the packing, and retention is thus the result of adsorption forces. The stationary phase is polar and is combined with a non-polar mobile phase. When HPLC first started to develop this was the most common type and thus this combination of polar stationary phase and non-polar mobile phase is referred to as **normal phase** (NP) HPLC. However the use of non-polar bonded phases with polar mobile phases then began to be introduced and thus these were referred to as **reversed phase** (RP) HPLC. Reversed phase HPLC has far outstripped the use of normal phase HPLC and is the most common type in use today.

Ion-Exchange Chromatography

In ion-exchange chromatography the stationary phase is a charged ionic group incorporated in the column packing. The mobile phase contains ions; ionic sample molecules compete with these for a place on the surface of the stationary phase. Both cation exchangers (e.g. NR_3^+, NH_3^+) and anion (e.g. SO_3^-, COO^-) exchangers can be used.

Size-Exclusion Chromatography

In size-exclusion chromatography the sample molecules do not interact with the stationary phase, the separation is based on the size of the sample molecules. Lower molecular weight species enter the pores on the column packing and higher molecular weight species do not fit in these pores and thus are not retained. A range of stationary phases in different pore sizes is available and the selection of an appropriate one for a given separation is based on the molecular weight of the sample. Stationary phases that are used with aqueous mobile phases are often referred to as gel filtration chromatography (GFC) packings, and those used with non-polar organic solvents are called gel permeation chromatography (GPC) packings. Size-exclusion HPLC is particularly applicable to high-molecular-weight species.

Preparative HPLC

Although HPLC is primarily used as an analytical tool it may also be used for preparative purposes. The stationary phase and mobile phase conditions used to separate and analyse a mixture of components for analysis may also be used for preparation of the components. Scaled up versions of the analytical HPLC systems are available. These allow large amounts of material to be separated and isolated.

Pharmaceutical Analysis

What is a drug?

A drug is a molecule that interacts with a biological system to produce a biological response which has a therapeutic effect. Drug molecules achieve this by attaching to a site in the body. The study of how drugs act when they enter the body is known as **pharmacology**. The structure of the drug molecule determines whether it will attach to a specific site, specifically the functional groups which are present and the location of these groups with respect to each other (the **stereochemistry** of the molecule). Drugs are very heterogeneous compounds, different drugs being related by the condition or disease which they are used to treat rather than their structure. The drug itself is called the **drug substance**, or **active pharmaceutical ingredient** (API). The structures of some well known drugs are provided in Table 2. These are all new chemical entities (NCEs), this means that they are new structures which have been created for use as drugs. The functional groups present in these drugs are listed, it is the arrangement of these groups that enables the molecule to achieve the therapeutic effect. The molecules in Table 2 are all of relatively low molecular weight and many contain basic groups. These are characteristics of many NCE pharmaceuticals.

Table 2   Common drug molecules

Drug	Structure	Molecular weight	Functional Groups
Amoxycillan *Antibiotic*		365	Amide Primary amine Thioether Lactam ring Carboxylic acid Phenolic hydroxyl group
Diazepam (Valium) *Sedative* *Anxiety relief*		285	Amide Tertiary amine Benzyl group Chlorobenzyl group
Atorvastatin calcium (Lipitor®) *Statin*		1155	Carboxylic acid Alcohol groups tertiary amine Fluorobenzyl group Benzyl groups Amide

Table 2   Common drug molecules

Drug	Structure	Molecular weight	Functional Groups
Tamoxifen *Anti-cancer*		372	Tertiary amine Ether Benzyl groups Alkene
Fluoxetine (Prozac®) *Anti-depressant*		309	Ether Secondary amine Trifluorobenzyl group Benzyl group
Oseltamivir (Tamiflu®) *Anti-viral*		312	Ester Ether Amide Primary amine Cyclic alkene

Biopharmaceuticals are drugs which are produced using biological systems, these are referred to as new molecular entities (NMEs). The growth of the **biotechnology** industry has lead to an increase in the development of peptide, protein and antibody based drugs. The first such substance approved for therapeutic use was recombinant human insulin (rHI, trade name Humulin®), which was developed by Genentech and marketed by Eli Lilly and Company in 1982. Another well known example is the humanised monoclonal antibody trastuzumab (trade name Herceptin®), also developed by Genetech and first marketed in 1998.

Drug delivery
Drug delivery is a term that refers to the delivery of the drug into the body. This is achieved using a dosage form, also referred to as a **drug product** or **formulated product**. The most common method of delivery is oral (e.g. tablets, capsules, liquids). Other non-invasive routes include nasal and rectal administration. Many drug products cannot be delivered using these routes because they might be susceptible to degradation or are not incorporated efficiently. In particular, protein and peptide drugs are often delivered by injection (e.g. immunisations which are based on the

delivery of protein drugs are often done by injection). The ingredients present in a drug product other than the drug, which are present to aid drug delivery, are referred to as **excipients**.

Analysing drug related samples by HPLC

Throughout the process of creating a new drug, developing it into a suitable product and manufacturing of that product to supply patients, pharmaceutical analysis is critical for providing information about drug related samples. The applications of HPLC throughout this process are discussed in detail in Chapter 5. '**Analyte**' is a term used to refer to the molecule of interest that is being analysed. When performing pharmaceutical analysis this will be a drug related molecule. It may be the drug itself, an impurity derived during synthesis or by degradation, a **metabolite** of the drug or an excipient used in the formulation of the drug product. These are just some examples of drug related analytes.

To compare the different types of HPLC described previously with respect to analysis of drug molecules it is useful to consider the analyte characteristics of molecular weight and **polarity**. The molecular weight of the analyte relates to its size. Polarity is where molecules have an uneven distribution of electrons, so that one part has a relatively positive charge and the other a relatively negative charge. The functional groups present in the molecule will determine the polarity. In general, the polarities of common organic functional groups in increasing order are:

Aliphatic hydrocarbons < olefins < aromatic hydrocarbons < halides < sulfides < ethers < nitro compounds < esters ≈ aldehydes ≈ ketones < alcohols ≈ amines < sulfones < sulfoxides < amides < carboxylic acids < water.

The relationship between these analyte characteristics and the type of HPLC suitable for analysis is shown in Figure 2. It can be seen that the different types of HPLC overlap and more than one type of HPLC may be used for a given analyte. In particular, partition HPLC overlaps with all the other types. The molecular weight and polarity of the drug molecules in Table 2 all fit into the partition area of the diagram in Figure 2. In practice the majority of NCEs encountered in pharmaceutical analysis can be analysed using reversed phase partition HPLC. Ion exchange HPLC may be used for the analysis of counter-ions, these are used when the drug substance is present as a salt (e.g. in Table 2, atorvastatin is in the form of a calcium salt). It may also be used for any other drug related samples where the analyte has a charge. The use of polymers as excipients in drug products is becoming more widespread and size exclusion HPLC may be used for analysis of these.

Biopharmaceuticals are also mostly analysed using RP partition HPLC, primarily in the determination of protein purity and protein identity. The intact protein may be analysed or the protein is broken up using a site specific **protease** and the cleavage products are analysed (**peptide mapping**). A fundamental limitation of RP partition HPLC is the **denaturing** properties of the hydrophobic stationary phase and eluting

---

solvents. For preparative purposes this limits the recovery of bioactive species. Ion exchange HPLC may be operated at physiological conditions, e.g. at neutral pH and with salt solutions, and is not denaturing, therefore it is widely used for purification of proteins. It may also be used for analysis and purification of peptides and nucleic acids. Size exclusion HPLC is used for estimations of protein molecular weight or aggregation state and also for purification.

Figure 2   Types of HPLC related to molecular weight and polarity of the component of interest

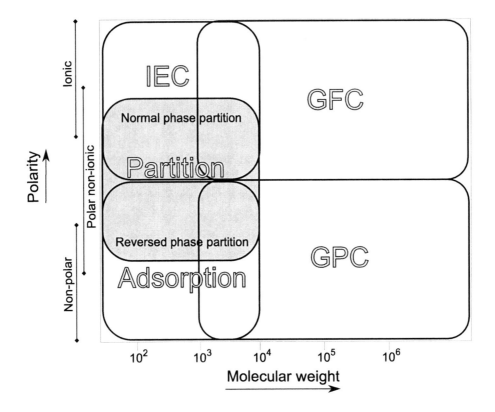

Summary

1. Chromatography is the separation of mixtures based on the differential migration of the components through a stationary phase when carried by a mobile phase.

2. HPLC is a type of liquid chromatography, which is used extensively by the pharmaceutical industry. It is characterised by the use of a finely divided stationary phase in a column.

3. There are four main types of HPLC, these are: Partition; Adsorption; Ion-exchange and Size-exclusion. The most common type of HPLC in pharmaceutical analysis is reversed phase partition HPLC.

4. Drugs are molecules which achieve a therapeutic effect when administered to the body. Pharmaceutical analysis involves the analysis of drug related molecules.

Further reading

Some useful general texts on HPLC:

M. McMaster, '**HPLC: A Practical User's Guide**', 2nd ed., Wiley, 2007.

Y. V. Kazakevich, R. LoBrutto, '**HPLC for Pharmaceutical Scientists**', WileyBlackwell, 2007.

V. Meyer, '**Practical High-Performance Liquid Chromatography**', 4th ed., Wiley 1999.

R.J. Hamilton, P.A. Sewell, '**Introduction to High Performance Liquid Chromatography**', Chapman and Hall, 1982.

L. R. Snyder, J. J. Kirkland, '**Introduction to Modern Liquid Chromatography**', 2nd ed., Wiley, 1979.

S. Lindsay, '**High Performance Liquid Chromatography**', Analytical Chemistry by Open Learning, Wiley, 1987.

LCGC Europe, (www.lcgceurope.com), Advanstar Communications
A free monthly magazine on chromatography.

Notes

The Stationary Phase

I t is the combination of a suitable stationary phase and mobile phase that enables the separation of a mixture and thus the analysis of the components in the mixture. In this chapter the stationary phase for HPLC is discussed. HPLC is characterised by the use of very small particles of stationary phase which are fixed in place in a HPLC **column**, often made of a material such as stainless steel. A typical column is shown in Figure 3.

Figure 3   Typical HPLC column

Normal phase and reversed phase HPLC

In order to describe the different stationary phases available for HPLC it is necessary to explain the concept of normal and reversed phase HPLC, which was introduced in Chapter 1. These types of HPLC vary due to the polarity of the stationary phase and mobile phase in each as shown in Table 3.

Table 3   Polarity of stationary phase and mobile phase used in normal phase and reversed phase HPLC

	Stationary phase	Mobile phase
Normal phase	Polar	Non-polar
Reversed phase	Non-polar	Polar

Normal phase HPLC

In a mixture of components to be separated those analytes which are relatively more polar will be retained by the polar stationary phase longer than those analytes which are relatively less polar. Therefore the least polar component will elute first. The attractive forces which exist are mostly **dipole**-dipole and hydrogen bonding (polar) interactions.

Reversed Phase HPLC

In a mixture of components to be separated those analytes which are relatively less polar will be retained by the non-polar stationary phase longer than those analytes which are relatively more polar. Therefore the most polar component will elute first. The attractive forces which exist are mainly non-specific **hydrophobic** interactions. The exact nature of these interactions is still under discussion[1].

Parameters to describe a HPLC column

The parameters used to describe a HPLC column refer to the nature, type and size of its contents, the dimensions of the column and the materials used in its construction. A list of parameters is detailed in Table 4.

Table 4   Parameters used to describe a HPLC column

Parameter	Description
Packing/matrix	The finely divided material with which the column is packed, usually silica. It can be used as the stationary phase in adsorption chromatography or a bonded phase is attached for use in partition chromatography.
Bonded Phase	The stationary phase is chemically bonded to the packing/matrix.
Particle size	The size of the particles in the column (if applicable), usually measured in microns.
Pore Size	The size of the pores in the particles/monolith, usually measured in angstroms.
Length	The length of the column, usually measured in cm or mm.
Diameter	The internal diameter of the column, usually measured in mm.
Hardware	The material used to construct the external tubing and end fittings of the column.
Manufacturer	The name of the manufacturer of the column.

Packing/Matrix of the HPLC column

Silica

The most common packing material used in HPLC columns is silica. It is physically robust and chemically stable in virtually all solvents and at low pH (it begins to dissolve around pH7). The manufacturing technology for silica production has improved substantially since the early days of HPLC. Irregular shaped particles contaminated with metal impurities have given way to spherical particles with low levels of impurities. This purer silica is known as type B silica and the less pure material is known as type A silica. Silica packed columns may also be produced as

monoliths, a single continuous piece of silica. All these improvements to the silica packing have resulted in better chromatographic performance.

A schematic representation of the chemical structure of silica and the silica surface is shown in Figure 4. Silica consists of silicon atoms bridged three-dimensionally by oxygen atoms. The different types of adsorption sites on the silica surface are shown, labelled (a) to (e). The free **silanol** (Si-OH) groups (a) are most active and most of the chromatographic properties of the silica surface are related to interactions with these. They are weakly acidic. Metals in the silica increases the acidity of neighbouring free silanols (b), the metal ion is shown as Me$^+$ in Figure 4. **Siloxane** bonds (c) and hydrogen bonded silanols (d) may also contribute to the surface activity. Geminal silanol groups (e) are relatively benign.

Figure 4   Silica structure and surface adsorption sites

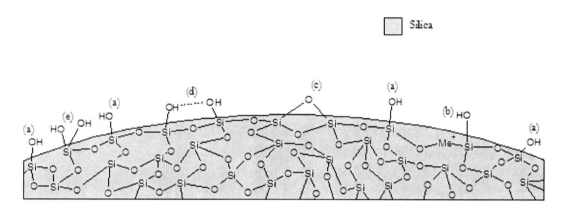

The packing in a HPLC column may either be the stationary phase itself, or, the packing may be a matrix which is used to bond to the stationary phase. Materials other than silica are used for packing columns and a discussion of these is included in the sections below.

The packing is the stationary phase

Silica is used as the stationary phase in adsorption HPLC, where it is combined with a non-polar mobile phase (i.e. normal phase). Columns packed with silica are available from a large number of manufacturers and each of these has a different propriety name to refer to their product (e.g. Spherisorb® silica by Waters™, Kromasil® SIL by Eka Chemicals™, ZORBAX® Rx-SIL by Agilent™). Other polar stationary phase packings include alumina and graphitised carbon. Alumina was common in the early days of HPLC but is rarely used now. Graphitised carbon is a more recent development and offers a possible solution for very polar analytes, and also for extended pH or high temperature applications.

The packing is a matrix to which the stationary phase is attached

Silica may be used as a platform to bond a stationary phase. It is easily derivatised to form bonded phases through attachment of reactive silanes to surface silanols. In this case it acts as a matrix to hold the bonded phase. The bonded phase may be polar (normal phase) or non-polar (reversed phase). Bonded phases are discussed in the next section.

Other materials that are used as the matrix in HPLC columns are polymers and zirconia. These materials offer greater stability both at high pH and also at high temperatures when compared to silica. This enhanced stability allows for the widest possible range of chromatographic conditions, desirable when developing new HPLC analytical methods for complex mixtures. Polymer based columns are not as common in partition HPLC as silica because they are generally not as efficient. (The **efficiency** of a column is a measure of how many components can be completely separated and is discussed in more detail in Chapter 7.) An example of currently available columns claiming efficiency similar to silica is the apHera™ range of reversed phase columns from Advanced Separation Technologies Inc (Astec). Polymer based packings are popular for ion exchange HPLC and size exclusion HPLC. Zirconia columns are growing in popularity but still account for a very small proportion of the total HPLC columns used in pharmaceutical analysis. Examples of commercially available zirconia based columns are Discovery Zr from Supelco® and also the ZiChrom® range, a company which specialises in zirconia HPLC products.

Recent advances in column design have introduced hybrid technologies. These incorporate inorganic silica and organic groups in the packing material resulting in material which is more chemically and thermally stabile than silica. Examples are the bridged ethylene hybrid material used by Waters in the Acquity UPLC™ columns and the Pathfinder® column from Shant Laboratories which has an internal core of silica surrounded by an organic polymer capsule.

Bonded phase

Reversed phase bonded phases

Reversed phase bonded phases are non-polar groups, these are attached to the matrix surface by covalent bonding. The most common bonded phases are alkyl groups. C_{18} is the most popular, also referred to as octyldecylsilane (ODS). C_{18} alkyl groups are added to a silica matrix by reacting an appropriate silane with the silanol groups on the silica surface. The silane will contain: a group that can react with the silica surface (typically Cl), side groups (e.g. methyl), and the C_{18} group, see Figure 5.

In Figure 5 the side groups are represented as 'R'. If R is also reactive it is possible to form polymeric phases, forming a tree like structure on one reacted silanol. However, monofunctional silanes (where R is non-reactive) result in more stable phases and the majority of commercially available bonded phases are monomeric.

Figure 5   Preparation of C_{18} bonded phase packing

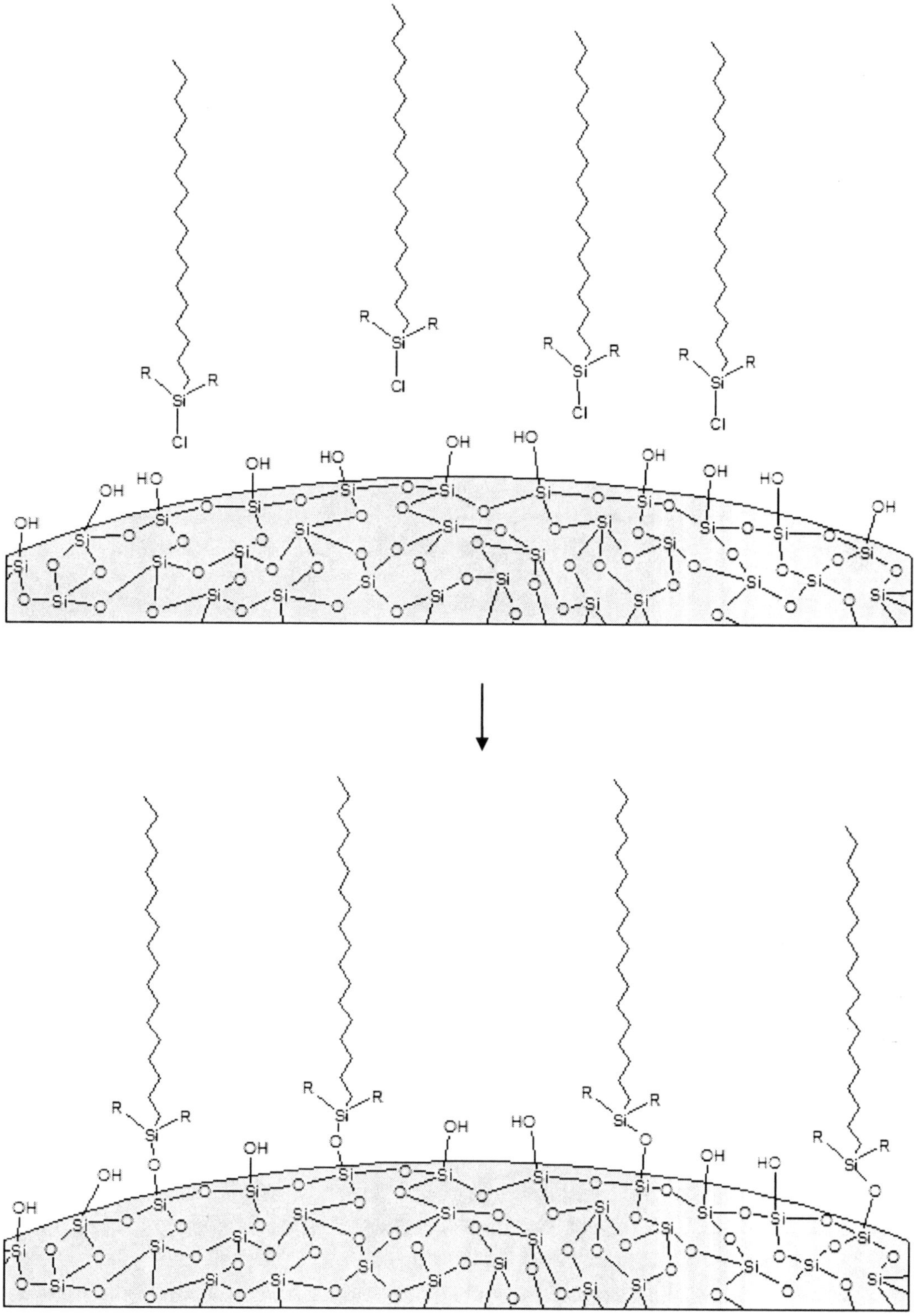

Other common RP bonded phase groups include other alkyl groups. Octyl groups (C_8) are the next most popular after C_{18}. Small alkyl groups such as C_1 and C_4 may also be used as bonded phases. These are less retentive for non-polar compounds due to the shorter chain length and are used for polar analytes and biomolecules. Phenyl group and cyano (CN) bonded phases may be used as an alternative to an alkyl group because they result in a different order of elution and separation for a mixture of components. This difference in the elution order and retention times on one column compared to another is referred to as the **selectivity** of the column. Components which are not separated on one phase may be separated on another.

It is important to be aware that columns from different suppliers which have the same bonded phase do not necessarily give the same results. They may have differences in selectivity, e.g. not all C18 columns will give the same separation for a given mixture of components. This is due to differences in the way both the silica and the bonded phase are manufactured. Suppliers use propriety names to distinguish particular column chemistries and these may be available in a number of bonded phases. Some common examples of RP bonded phase columns are given in Table 5.

Table 5   Common RP bonded phase columns

Name of column packing	Some available RP bonded phases	Manufacturer
Symmetry®	C_1, C_4, C_8, C_{18}, Phenyl, CN	Waters
Betabasic™	C_4, C_8, C_{18}, Phenyl, CN	Thermo™
Luna®	C_5, C_8, C_{18}, Phenyl-Hexyl, CN	Phenomenex
Kromasil	C_4, C_8, C_{18}	Eka Chemicals

During the formation of the bonded phase on silica not all the silanols on the silica surface are used up, as is shown in Figure 5. The presence of the residual silanol groups presents a problem in reversed phase HPLC, particularly when analysing bases. At pH < 7, silanols and many basic compounds are partially ionised. The interaction of these causes peak **tailing** for basic analytes, an effect which makes quantification difficult and can reduce the number of components in a mixture which can be effectively separated.

The term 'base deactivation' was introduced to describe processes during the manufacture of HPLC columns which try to eliminate the effects of residual silanols. Improvements in column technology included the introduction of **endcapping**, a process which reacts the residual silanols with a small reactive silane, e.g. trimethylchlorosilane, see Figure 6. The technique of endcapping results in much better performance when analysing basic molecules. Not all the silanols can be endcapped, due to steric hindrance, but using this process accessible silanols are prevented from interacting with analyte molecules.

Figure 6   Endcapping of silanols

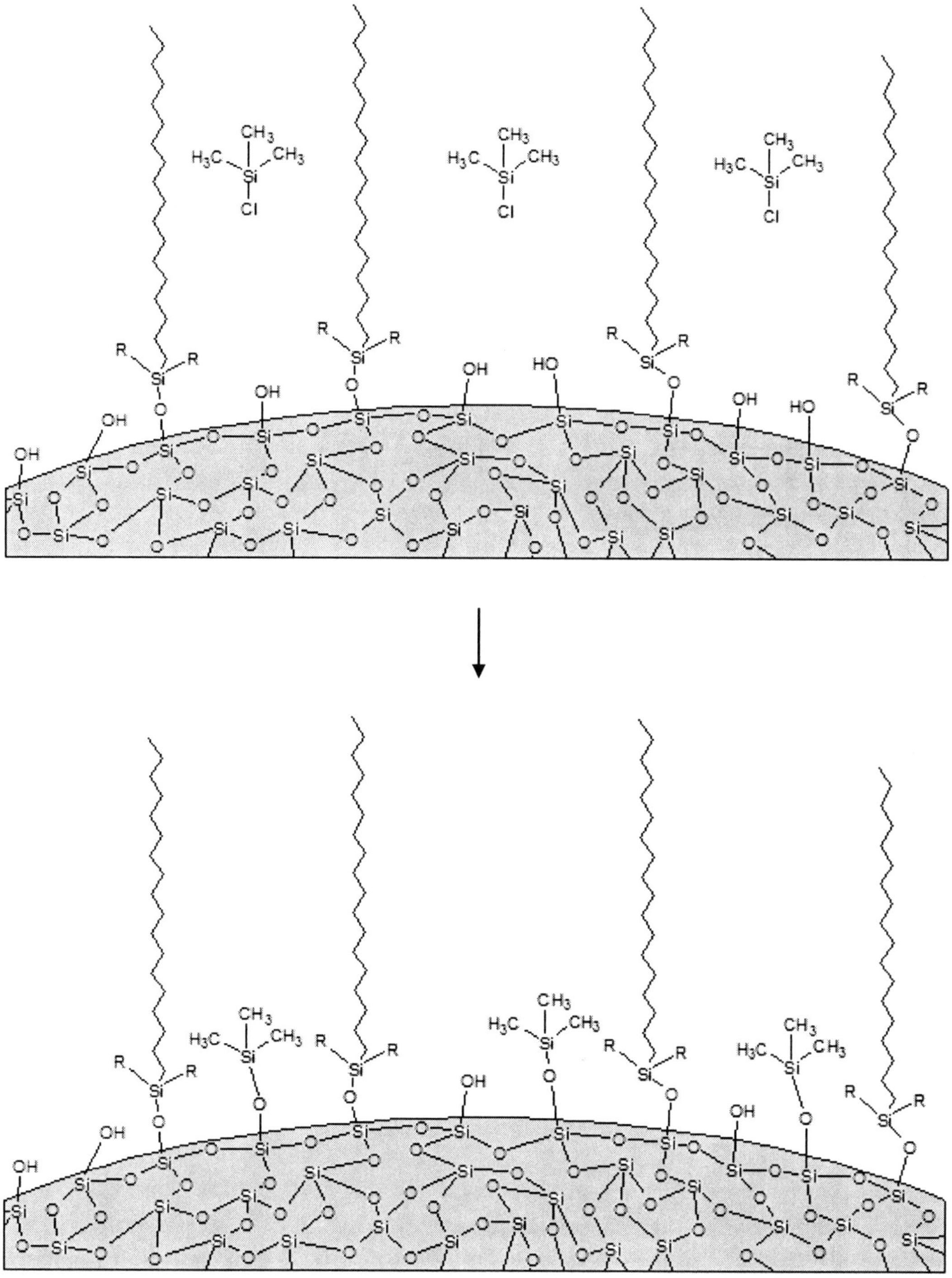

It has already been mentioned that columns which have good chemical and thermal stability enable the widest possible range of chromatographic conditions. Silica based columns have been developed to overcome specific problems encountered during reversed phase HPLC.

Chemical stability can be improved by using bulky alkyl side groups (e.g. isobutyl groups) as 'R' in Figure 5. Due to the steric hindrance they create, the silica surface is protected from chemical attack and this results in a more stable stationary phase.

Analytes which are very polar will not be retained on a typical non-polar bonded phase such as C_{18}. They will favour the polar mobile phase and thus travel straight through the stationary phase. A number of column technologies have been developed to enable analysis of these compounds using reversed phase HPLC. Aqueous reversed phase mobile phases are generally more convenient than 100% organic mobile phases and thus are favoured in pharmaceutical analysis whenever possible.

For some polar analytes it may be possible to obtain some retention if the mobile phase is 100% aqueous[2]. However many columns will suffer a 'phase collapse' and cannot be used under these conditions. So called 'Aqua' columns have been developed to deal with this situation and the solution employed varies between different manufacturers. Polar endcapping (1), long chain alkyl phases (2) and polar embedded groups (3) are just some solutions.

1. Polar endcapping is similar to the endcapping process described previously (see Figure 6) but instead of trimethylsilanes a short chain hydrophilic endcapping chemical such as trimethoxysilane is used. These hydrophilic chemicals allow the silica surface to be wetted with water and allow the full interaction with the longer alkyl chains.

2. Long chain alkyl phases, such as C30 are resistant to phase collapse under highly aqueous conditions.

3. Polar embedded groups incorporate a polar functional group (e.g. amides, ethers etc.) in the alkyl chain close to the surface of the silica (see Figure 7). This allows interaction with polar analytes and wetting of the surface.

These solutions (and others) are also used to enhance the selectivity for polar compounds without necessarily working in 100% aqueous conditions. In pharmaceutical analysis, it is often necessary to separate mixtures of polar and non-polar compounds and a single technique is preferred.

Often the actual processes which are used to make columns: 'suitable for polar analytes'; 'suitable for 100% aqueous conditions' etc. are described by the manufacturer as 'using propriety reagents and procedures' because this information is considered a trade secret. The information given above is intended to give a general understanding of how the chemistry of reversed phase bonded phases can be altered to make them suitable for the range of analyte types that may be encountered in pharmaceutical analysis.

Figure 7 Polar embedded bonded phase (shown with endcapping)

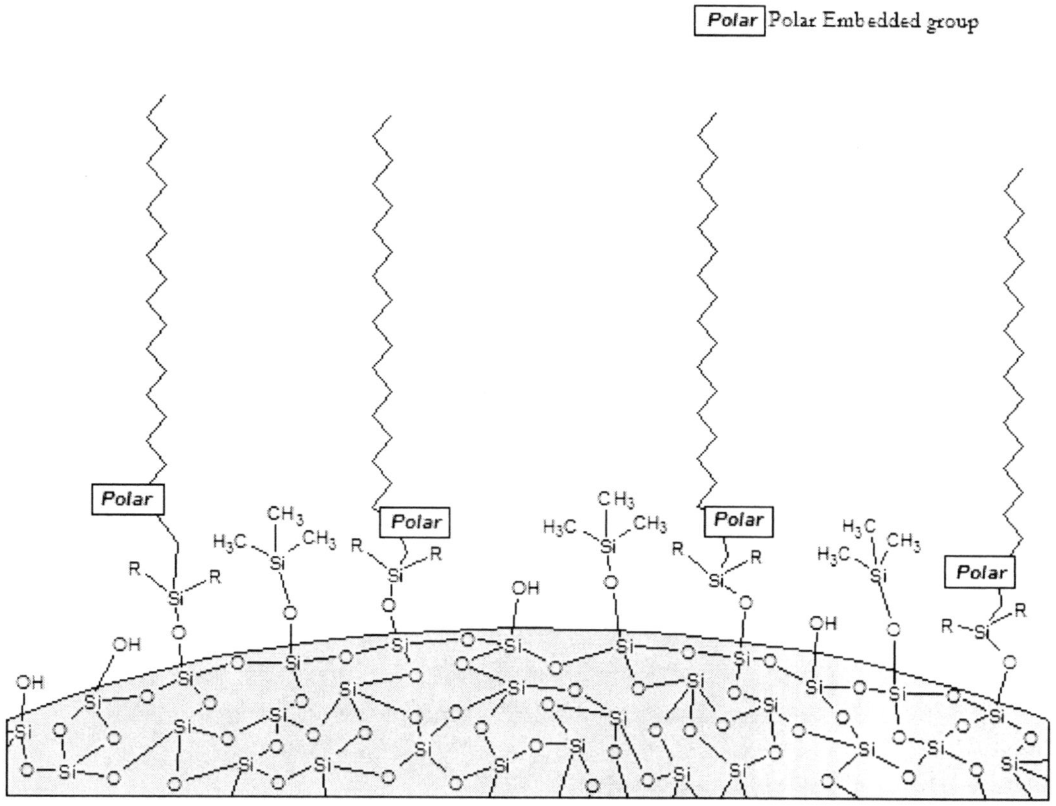

Other reversed phase bonded phases have been developed to provide alternative selectivity. Bonded phases which incorporate fluorinated groups have gained popularity in recent years, these have different selectivity to C_{18} columns but within a similar retention time window[3].

A few examples of the different types of bonded phases discussed above are presented in Table 6. These are just a few from the wide range commercially available.

Table 6 Examples of bonded phases developed for specific problems

Name	Manufacturer	Application
Synergi® Hydro-RP	Phenomenex	Aqua column
Prevail™	Alltech	Aqua column
Polaris®	Varian, Inc	Polar embedded phase
Acclaim® Polaradvantage	Dionex	Polar embedded phase
Develosil® Combi-RP C30	Nomura Chemical Co.	Long alkyl chain phase
FluoroSep-RP	ES Industries	Fluorinated column

Normal phase bonded phases

The bonded phases that are used for normal phase chromatography are polar. Typical examples are cyano (CN), amino (NH2) and diol bonded phases. These phases may also be used with reversed phase solvents for a small number of applications.

Hydrophilic Interaction Chromatography (HILIC) is a variation of normal phase chromatography. A polar stationary phase is combined with a mobile phase containing a high concentration of a non-polar solvent and a low concentration of polar solvent (aqueous). The order of elution is similar to normal phase, the least polar analyte will elute first and polar analytes will be retained longer. HILIC is also referred to as 'aqueous normal phase' or 'reverse reversed phase'. This type of chromatography is used for very polar, water soluble analytes such as peptides and metabolites. These are difficult to retain by reversed phase chromatography.

Bonded phases for ion-exchange HPLC

The bonded phase for ion exchange chromatography consists of an ionic group which can provide sites for exchange. These exist in cationic and anionic forms, the selection depending on the analyte. Four general types are available, strong and weak cation exchangers and strong and weak anion exchangers. In pharmaceutical analysis ion exchange phases may be used for biomolecules, amino acids, peptides, proteins or nucleic acids. Silica based and polymer based columns are commonly used.

Bonded phases for size-exclusion HPLC

Silica based size-exclusion chromatography packings have chemicals bonded to the surface to deactivate the silica but these bonded phases do not interact with the analyte molecules since the separation is based only on size. The details regarding polymer based size-exclusion chromatography packings are usually propriety information that is not commonly disclosed by the manufacturers.

Bonded phases for biomolecules

The bonded phases described above may be used for biomolecules. In addition there are columns based on these phases which are specifically designed for biomolecules. In particular a range of pore sizes is often available. Pore size is very important in the analysis of proteins, larger pore sizes (typically ~300Å) enable the protein to go inside the pore and interact with the bonded phase.

Chiral Stationary Phases (CSP)

A large number of drugs contain **chiral** centres and often only one **enantiomer** is pharmacologically active. Therefore an analytical technique which will separate the enantiomers is required. To use HPLC for this separation involves the use of a chiral stationary phase. A large number of different phases are available and the selection of a suitable one will depend on the analyte. Chiral HPLC is usually operated in normal phase mode but some columns are used in reversed phase mode. This specialised area of HPLC is outside the scope of this book.

Particle size

The development of finely divided stationary phase resulted in the development of HPLC as an analytical technique. The magnitude of these particles sizes is typically in the range 1.5 to 10 microns. Most packings have a distribution of particle sizes, this is inherent in the manufacturing process. A narrow distribution gives better results and thus the high performance columns tend to have narrow distributions of particle size.

Over the past few years the technology required to both produce and utilise sub 2 micron particles has been commercialised. Previous to that the range of particle sizes was typically 3, 5 and 10 microns. Reduction in particle size gives benefits in efficiency (refer to Chapter 7 for more detail on efficiency) and faster analysis but also results in higher operating pressures, this means that instrumentation is required which can operate at ultra high pressure[4]. Columns which are packed with sub 2 micron material include: ACQUITY UPLC™ BEH columns from Waters which use a 1.7 micron particle size, Hypersil GOLD columns from Thermo which use a 1.9 micron particle size and VisionHT™ columns from Grace® which use a 1.5 micron particle size. The majority of pharmaceutical analysis by HPLC is currently performed on the relatively larger particle sizes but the use of small particles and ultra high pressure systems appears likely to grow very quickly in the next few years.

Monolithic columns are made from a single particle and thus the concept of particle size is not relevant for these columns. They can be used with higher flow rates of mobile phase relative to columns packed with particles, due to the lower backpressure they produce, and thus can be used to speed up analyses.

Pore size

Most particles used in HPLC are fully porous. Typical pore sizes for analytical HPLC range from approximately 60 to 300Å. General purpose packings have a pore size of about 100Å and particles designed for the analysis of large molecule, e.g. proteins, have a pore size of 300Å or greater. In size exclusion chromatography the pores are needed to effect the separation.

Length

The length of the column influences the extent of the separation of the components in the mixture. A longer column enables more separation. However, simply increasing the column length will not achieve the perfect separation. The time taken to perform the separation will increase as the length increases since it takes longer to travel through the column. Also as the column length increases, the backpressure experienced will also increase. Another problem with increasing the column length is that the spread of the components of the mixture in the column will increase, this will result in difficulties during detection and quantification.

Figure 8 shows the effect of doubling the length of stationary phase used for the example previously described in Figure 1. Components A and B encounter the stationary phase at time t1. At time t2 they are separated and component B has travelled halfway through the stationary phase. At time t3 Component B elutes from the stationary phase further separated from component A, but both components are beginning to experience **band broadening** due to diffusion effects. This means that the component molecules become more spread out as they elute from the column than they were when they first encountered the column.

Figure 8 Effect of column length on the separation

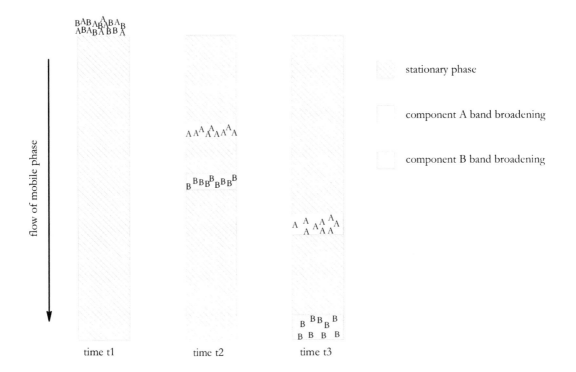

Typical column lengths are 5, 7.5, 10, 15 and 25 cm. Shorter columns are used for fast analysis in high throughput applications, these may be between 1 and 3 cm in length. For a given application the column length is selected which gives adequate separation but does not result in band broadening and a very long analysis time.

Diameter

The internal diameter (i.d.) of the column determines the scale of the HPLC being performed. The nomenclature that may be encountered when describing the scale[5] of HPLC is detailed in Table 7. The rate of flow of the mobile phase decreases with the internal diameter. Larger internal diameters ($\geq$ 10cm) are used for preparative HPLC.

Normal-bore and narrow-bore columns are used for most applications which require quantification in the development and manufacturing of pharmaceuticals. The technology associated with the smaller internal diameter columns (microbore and below) has developed very quickly over the past few years driven by the requirements of bioanalytical separations and in particular, **proteomics**.

Table 7 Nomenclature for scale of HPLC

Scale	Internal diameter
Normal-bore	4 mm to 5 mm
Narrow-bore	2.1 mm to 4 mm
Microbore	1 mm to 2.1 mm
Capillary-bore	100 μm to 1 mm
Nanobore	25 μm to 100 μm

Hardware

The term hardware here refers to the tubing encasing the packing and the end-fittings. It is desirable for the material used to be strong and inert. For the majority of HPLC columns the hardware is constructed using stainless steel. However, it also possible to obtain columns made from polyetherether ketone (PEEK). This may be preferred if it is suspected that there is an interaction between the sample and stainless steel as in the analysis of some biological systems.

Columns are available in two different systems: Columns and Cartridges. Columns have complete end-fittings and connect directly to the HPLC system. Cartridges are simply the packed tubing with filters at the end that contain the packed bed. The end fittings for cartridges are separate pieces which are attached when required. There are no performance differences between the two systems so the choice of which to use is made on personal preference.

The end-fittings enable the column to be connected to a HPLC system via suitable tubing. The connection is secured by threading on the inside of the end-fitting which tightens on a nut attached to the tubing. Contained inside the end-fitting is a frit which retains the packing in the column but allows the flow of mobile phase through the column. The pore size of the frit needs to be smaller than the particle size of the packing to prevent it escaping or clogging the frit. Typical pore sizes for frits are 0.5 μm and 2 μm. If there is particulate matter in the mobile phase this may be retained on the frit and build up over time reducing the life time of the column.

Manufacturer

There are many manufacturers of HPLC columns leading to a wide selection of columns available for purchase. There are those who manufacture silica (or other packings/matrix), those who attach bonded phases and those who buy the packing already made and pack columns only. Some of the bigger producers in the marketplace do all three. Just a few examples of manufacturers are Merck, Waters, Phenomenex, Thermo and Supelco.

Guard columns

A guard column is a short HPLC column which is installed prior to the analytical column so that the analytes and mobile phase flow through it before they flow through the analytical column. Its purpose is to protect the analytical column from particulates that may be present in the sample or the mobile phase and also to remove strongly absorbed sample components. These would be detrimental to the performance of the analytical column and would shorten its lifetime. The guard columns are less expensive and with regular replacement should protect the more expensive analytical column. The packing in the guard column will usually match the material in the analytical column. They are available in a column format and also in cartridge systems which attach directly into the analytical column.

Selecting a stationary phase

Manufacturers of HPLC columns are constantly introducing new stationary phases, claiming their technology is superior to competitors, and adding to the already numerous phases available. Each manufacturer may have a range of column packings, and each packing may be available in a range of column chemistries. Therefore, the selection of a suitable column for an analysis of a mixture of components is very difficult. This decision is made during HPLC **method development**, the process of selecting a suitable column and mobile phase for a given separation. Before embarking on method development, suitable experience of using the technique of HPLC is required. Useful texts on this topic are available[6]. The most suitable combination of length, internal diameter and particle size will be selected during the HPLC method development. The best choice will depend on both the nature of the analyte and the aim of the analysis.

Extended pH operating range

An extended pH range offers extra chromatographic conditions when analysing acids and bases. When using silica based columns a method of protecting the silica is required since it will dissolve at pH greater than seven and, at very low pH, hydrolysis of the siloxane bond will occur, leading to loss of the bonded phase. Hybrid technologies, previously discussed, provide a protection method as do techniques which utilise specialised endcapping methods. Polymer and zirconia based columns are suitable for extended pH analysis.

Extended temperature operating range

High temperature HPLC is currently an area of much investigation and commercial development. Columns are being developed that are stable even up to 200 degrees Celsius. The advantage of high temperatures is that separations are more efficient and analysis times shorter. (Refer to Chapter 7 for more detail on efficiency.) Columns based on silica hybrid technology, zirconia based columns and polymer based columns are suitable for high temperatures[7].

Summary

1. In HPLC the stationary phase is referred to as the column.

2. The parameters which can be used to describe a column are: packing/matrix; bonded phase; particle size; pore size; length; diameter; hardware and manufacturer.

3. Silica is the most common packing/matrix in HPLC columns.

4. Bonded phases provide a wide range of different selectivities for the separation of mixtures. The most common bonded phase is C18.

5. The most suitable combination of bonded phase, length, internal diameter and particle size and pore size is selected during the method development and depends upon the application.

References

1. L.R. Snyder, J.W. Dolan and P.W. Carr, *J. Chromatogr.* **A1060**, 77-116, 2004, 'The hydrophobic-subtraction model of reversed-phase column selectivity'.

2. R.E. Majors, M. Przybyciel, *LC•GC Eur.*, **15**, 780-786, 2002, 'Columns for Reversed-Phase LC Separations in Highly Aqueous Mobile Phases'.

3. M. Przybyciel, *LC•GC Eur.*, **19**, 19-27, 2006, 'Fluorinated HPLC Phases – Looking Beyond C18 for Reversed-Phase HPLC'.

4. R.E. Majors, *LC•GC Eur.*, **19**, 352-362, 2006, 'Fast and Ultrafast HPLC on sub-2 μm Porous Particles — Where Do We Go From Here?'

5. G. Rozing, Recent Developments in LC Column Technology, *LC•GC Eur.*, **16**(6a), 14-19, 2003, 'Trends in HPLC Column Formats – Microbore, Nanobore and Smaller'

6. L.R. Snyder, J.J. Kirkland, J.L. Glajch, 'Practical HPLC Method Development', 2nd ed., Wiley 1997.

7. Y. Yang, Recent Developments in LC Column Technology, *LC•GC Eur.*, **16**(6a), 37-41, 2003, 'Stationary Phases for High-Temperature Liquid Chromatography'.

Further Reading

U. D. Neue, 'HPLC Columns: Theory, Technology and Practice', Wiley, 1997. An excellent text on all aspects of HPLC columns including the underlying principles of column technology.

LCGC Europe, (www.lcgceurope.com), Advanstar Communications
A regular feature entitled 'Column watch' reviews recent developments in column technology.

The websites of HPLC column manufacturers contain useful information about their products, these may be found easily using internet search engines.

Notes

The Mobile Phase

The mobile phase for HPLC is the liquid phase which is continually flowing through the stationary phase and which carries the analyte through with it. The composition of the mobile phase which is used is dependent on both the stationary phase and the nature of the compounds being analysed. The different properties of solvents define whether they are suitable for use as a mobile phase either under reversed phase or normal phase conditions.

Solvents

The most common solvents used for HPLC are listed below in order of increasing polarity:

n-hexane
methylene chloride
chloroform
methyl-t-butyl ether
tetrahydrofuran (THF)
isopropanol (IPA)
acetonitrile (MeCN or ACN)
methanol (MeOH)
water

A blend of two (or more) of these solvents is used as the mobile phase in a HPLC analysis. The proportions of the different solvents in the blend act to adjust the polarity of the mobile phase. This is combined with a suitable stationary phase to achieve the separation of a mixture. Ideally, the components in the mixture will be separated fully and will all elute within a practical time scale.

By convention, chromatographers usually refer to the strong solvent in a mobile phase as the 'B' solvent and the weak solvent as the 'A' solvent. Generally, solvent strength is related to polarity, with non-polar solvents being 'strong' solvents for reversed phase HPLC and polar solvents being 'strong' for normal phase HPLC.

A binary mixture is a mixture of two solvents and is the most common type of mobile phase. However ternary mixtures, where three solvents are blended, are also used. The choice of the solvents in the mobile phase, and the proportions of each, will be selected during method development.

The most important property of the solvent is its ability to interact with both the stationary phase and the analytes in the mixture, resulting in the desired separation. However, there are other important properties that need to be considered. An ideal solvent will be readily available in high purity, relatively inexpensive, safe to use routinely, and compatible with the entire HPLC system including the detector.

Reversed phase HPLC solvents

In reversed phase HPLC the solvents used for the mobile phase are those towards the end of the list on the previous page, which are relatively more polar. Water is always used together with an organic solvent which is miscible with water in all proportions. Increasing the proportion of the organic solvent (%B) in the mobile phase will reduce the retention time of the analyte. This is because the analyte will usually be more soluble in the organic solvent and therefore will spend more time in the mobile phase thus reducing the time spent on the stationary phase.

'Rule of 3'

As a general rule for small molecules in reversed phase, an increase of the organic solvent component in the mobile phase of approximately 10% will result in the retention time being reduced by a factor of 3.

The two most common organic solvents, which may be combined with water to prepare a mobile phase, are acetonitrile and methanol. These are water miscible and have good properties (i.e. readily available, safe to use and compatible with HPLC systems.) Of these acetonitrile is usually the first choice, it has lower viscosity and lower **UV cutoff** than methanol (refer to chapter 4 for a full discussion of UV cutoff and UV detectors). Tetrahydrofuran (THF) may occasionally also be used, but only as a last resort, since it degrades to form peroxides, it has a high UV cutoff, it results in high backpressures and it reacts with PEEK fittings.

A combination of water with each of these three solvents may result in different separations for a given mixture of analytes. This difference in selectivity is primarily based on the polar characteristics of the solvents. The solvent selected for the separation will be the one that gives the best separation. If the separation cannot be achieved using a binary system, i.e. one organic solvent combined with water, then intermediate selectivity may be obtained by blending binary systems. The resulting mobile phase will be a combination of three components, a ternary blend.

Normal phase HPLC solvents

In normal phase HPLC the solvents used for the mobile phase are those towards the top of the list on page 29, which are relatively less polar. Hexane is often used as the 'A' solvent with isopropanol, methylene chloride or methyl-t-butyl ether as the 'B' solvent.

Mobile phase pH

Acids and bases are ionisable compounds. If the analyte of interest contains acidic or basic groups then the pH of the mobile phase will determine whether they are present in an ionised or non-ionised form during the separation. An ionised molecule will interact differently with the mobile phase and stationary phase during analysis than a non-ionised molecule. Ionised molecules are hydrophilic and non-ionised molecules are hydrophobic, therefore ionised molecules will be less strongly retained in reversed phase HPLC. The pH needs to be controlled if ionisable compounds are present in the sample. When pH is increased acids (HA) become ionised ($A^- + H^+$). Bases (B) become ionised (BH^+) as pH is decreased. The relationship between pH and retention is shown in Figure 9 for acids and in Figure 10 for bases.

Figure 9   The effect of pH on retention for acids

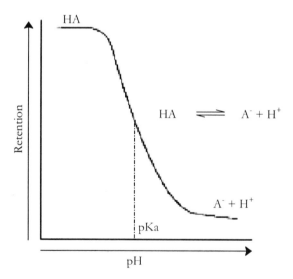

The pKa is the pH at which the acidic or basic group is half ionised and half non-ionised (indicated in Figure 9 and in Figure 10). Over the range of pKa ± 2 pH units, small differences in pH can result in large differences in retention. This pH effect may be used to obtain a separation but if the pH of the mobile phase is in this region it will be critical to ensure that it is exactly the same each time it is prepared for an analysis. Usually it is best to use a pH which is outside the range of pKa ± 2 pH units, since the retention will be less susceptible to small differences in pH. The pH will have no effect on the retention time of neutral compounds.

Figure 10 The effect of pH on retention for bases

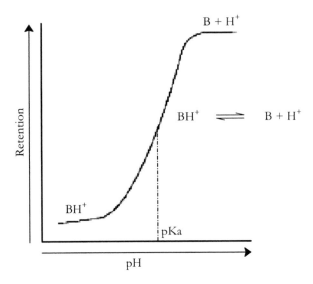

Buffers

The pH of a mobile phase is controlled by the use of a buffer. There are a large variety of buffers that can be used to control pH. Examples of buffers commonly used in HPLC are given in Table 8. The pH of the mobile phase should be within ±1 pH unit of the buffer pKa for good pH control.

Table 8   Examples of buffers used in HPLC

Buffer	pKa	Buffer	pKa
Trifluoroacetic acid (TFA)	0.3	Carbonate, pK_1	6.4
Phosphate, pK_1	2.1	Phosphate, pK_2	7.2
Citrate, pK_1	3.1	Tris-(hydroxymethyl)aminomethane	8.3
Formate	3.8	Ammonia	9.2
Citrate, pK_2	4.7	Carbonate, pK_2	10.2
Acetate	4.8	Triethylamine	11.0
Citrate, pK_3	5.4	Phosphate, pK_3	12.3

Buffer preparation

It is important to use high-grade reagents when preparing buffers for use in a HPLC mobile phase. Many common buffer reagents are available in a HPLC grade and convenient pre-prepared buffer solutions for use in HPLC are also available. Buffer concentrations in the range of 10 to 50mM are typical. To calculate how much buffer reagent needs to be weighed to obtain a buffer strength quoted in millimoles (mM), multiply the strength in mM by the molecular weight of the buffer reagent (usually printed on the label) and divide by 1000 (to convert mM into M). The result is the amount in grams required for 1 litre of buffer. An example is given in Figure 11. If the reagent exists as a hydrate, then use the molecular weight of the hydrate in the calculation.

Figure 11 Calculation for preparing a buffer

To prepare 25mM potassium phosphate buffer:

Buffer strength required	= 25mM
Molecular weight of potassium phosphate (K_2HPO_4)	= 174.18

Therefore, $25 \times 174.18 \div 1000 = 4.35$

4.35 g required for 1 litre of buffer

Other ways the buffer strength may be quoted are as a '%v/v' (% volume/volume) or '%w/v' (% weight/volume). To prepare a buffer of strength 1%v/v, 1mL is made up to 100mL with water. To prepare a buffer of strength 1%w/v, 1g is made up to 100mL with water. Different amounts or strengths can be prepared by scaling up or down as necessary.

The pH of the buffer is always checked using a pH meter and adjusted if necessary before mixing with the organic solvent component of the mobile phase. The adjustment is made using a suitable acid or base, e.g. When preparing a buffer from potassium phosphate use orthophosphoric acid or ammonium hydroxide as required to alter the pH to the desired value. Care should be taken not to contaminate the buffer during pH measurement.

Acid Modifiers

It is common to employ trifluoroacetic acid (TFA) and acetic acid alone to control pH at low values. This method of pH control does not provide a buffered mobile phase and may not be as effective for all types of samples, especially basic ones. However, it is popular for adjusting the pH of mildly ionisable compounds such as peptides and proteins. Concentrations of acids of approximately 0.05 to 0.1 %v/v are typical.

Other mobile phase additives

Amine modifier

A tailing suppression reagent, such as triethylamine (TEA), may be added to the mobile phase to reduce peak tailing (peak tailing is discussed in more detail in Chapter 7). This hydrophobic amine will bind to the unbonded silanols thus reducing the access of basic analytes to these. Other amines may also be used.

Ion pairing reagents

These are used for very hydrophilic analytes which can be difficult to retain using reversed phase HPLC. An ionic surfactant is added to the mobile phase, the hydrophobic part tends to stick to the column surface leaving the ionic portion in contact with the mobile phase. The ionic portion thus retains the hydrophilic analyte. It can take a long time for the ion pairing reagent to reach an equilibrium with the column, leading to very long analysis times. This problem means that it is used as a last resort and only if reversed phase HPLC cannot be used. The most commonly used reagents are alkyl sulfonates, e.g. dodecylsulfonate, heptanesulfonate (for cations) and tetraalkyl quaternary salts, e.g. tetramethyl/tetrabutyl ammonium sulfate (for anions).

The conversion of quoted strengths into amounts to weigh during preparation for these additives is the same as for buffers. Many additives are available as HPLC grade reagents and their use is recommended for the best results.

Isocratic and gradient elution

Isocratic elution is where the composition of the mobile phase remains constant throughout the analysis. **Gradient** elution is where the composition of the mobile phase is altered throughout the analysis. The proportion of the strong solvent B (the organic solvent in reversed phase mode) increases over the course of the separation. Isocratic elution is simpler to perform but for some mixtures of analytes, which differ widely in hydrophobicity, using isocratic elution to separate them would result in an extremely long run time. Suitable gradients for a given mixture of analytes are determined during the method development process. They are presented in terms of the increase in the proportion of the strong solvent B, e.g. 20 to 80 %B over 20 minutes. The increase in the proportion of solvent B over time is linear.

When using normal phase HPLC isocratic elution is used with silica columns since gradient elution is not possible. Gradient elution can be used with bonded phases.

Mobile phase preparation

Solvent grades

There are HPLC grade solvents available for the common solvents used and it is recommended that these are used for the best results. They are typically $\geq 99.9\%$ pure. Within the HPLC grade there are different types available from some suppliers. These include 'gradient grade', and 'LC-MS'. When running gradient elution and also when using a mass spectrometer as a detector the solvent used needs to be very pure to eliminate interferences which might affect the results. Acetonitrile is also available in a 'far UV' grade, this relates to the use of a UV detector in the far UV range, i.e. below 200nm. The difference in these grades relates to purity, very high purity solvents are usually obtained by multiple distillations. If a specialised grade is required

---

for an analysis then this information should be included in the details of the **HPLC analytical method**.

Water for HPLC may be sourced from solvent suppliers in HPLC grade bottles or some type of water purification system can be used[1]. A water purification system for HPLC should generate ultra pure water (18 MΩ resistivity), it is extremely important that the system is well maintained.

Different types of methods have different tolerances to the grade of the solvents used in the mobile phase. This should be assessed thoroughly during the method development process and any requirements included in the analytical method.

Measuring solvents

When measuring out the solvents to be used in a mobile phase each solvent is measured separately using a measuring cylinder of an appropriate size rather than measuring one and making up to volume with the other. The reason for this is that the volume of the mixture is smaller than the individual volumes due to a contraction effect. This is particularly applicable to methanol but is true for other water miscible solvents to some extent.

Mixing the mobile phase

When the mobile phase contains added buffer, ion-pair reagent or other additive, it is customary to weigh out these substances using HPLC grade reagents and add them to the aqueous portion of the mobile phase mixture. The pH is then adjusted on the aqueous portion prior to mixing with the organic portion. There are a number of options about how to perform the mixing of the mobile phase:

Isocratic methods

The aqueous and the organic portions of the mobile phase may be measured separately and mixed in a container by the analyst (premixing) or each portion may be placed on the HPLC system and mixed in the correct proportions by the HPLC pump (online mixing).

Gradient methods

The aqueous and the organic portions of the mobile phase for both the starting conditions and for the final conditions of the gradient may be measured separately and mixed in a container by the analyst (premixing). This will result in two mobile phases, e.g. for the gradient 20 to 80 %B over 20 minutes, a mobile phase of 20 %B and a mobile phase of 80 %B are prepared by the analyst, these are then placed on the HPLC system and mixed during the analysis by the HPLC pump. The other option, as in the isocratic method, is to place the aqueous and organic portions on the HPLC system and mix in the correct proportions by the HPLC pump (online mixing).

There are advantages and disadvantages to each approach. Premixing means that the pump does not have to perform mixing in the isocratic case and performs the least amount of mixing in the gradient case. If the pump being used is not capable of this type of mixing then the premixing option is preferred. Premixed mobile phases are subject to analyst error and different batches will be slightly different to each other. Online mixing removes the analyst error and is often more convenient to operate. Bottles received from the supplier containing solvent and premade reagent can be placed directly on the system preventing contamination associated with measurement. Plastic coated bottles are available for this purpose. However, if the pump is unable to deliver a well mixed mobile phase then online mixing is not suitable. The final decision depends on the pump capabilities, the analyst preference and the application.

Filtration

Ideally all HPLC solvents should be filtered through a 0.45 μm filter before use. This will remove particulate matter which could damage the HPLC system or the column and will result in better performance and longer lifetimes for consumable parts in the HPLC system and for the column.

Degassing

It is important that mobile phases do not contain air bubbles since these can cause problems in the HPLC system and in particular the detector. Dissolved gases can be removed by the following methods:

1. **Vacuum degassing**

   This is an online technique and most modern HPLC instruments include a degassing module. The mobile phase flows through a membrane which allows the gases to penetrate into the surrounding vacuum.

2. **Vacuum filtration**

   Performed off-line, this method has the advantage that the mobile phase is filtered simultaneously. Apparatus for vacuum filtration is available from a number of suppliers. It is connected either to a water aspiration line or to a vacuum. Typically 0.45 μm filters are used.

3. **Sonication**

   Sonication can be used for degassing by placing the mobile phase container in an ultrasonic bath for one minute.

4. **Helium sparge**

   Another online technique, this involves bubbling helium through the mobile phase. Helium is much less soluble than other gases. Initially the sparge will be at a level to displace the dissolved gases and then low levels will be introduced throughout the analysis to prevent re-adsorption. This technique requires a source of helium.

Often a combination of the above methods is used.

Storage of mobile phases

The expiry date of a mobile phase will be decided by the composition and the local procedures in the laboratory where it is being used. Buffer solutions are best prepared fresh each time they are required to ensure that the pH is correct and that there is no microbial growth.

Summary

1. A blend of solvents is used as the mobile phase for HPLC.

2. The pH of the mobile phase needs to be controlled using buffers for the analysis of ionisable compounds.

3. In isocratic elution the composition of the mobile phase is held constant during analysis.

4. In gradient elution the composition of the mobile phase changes during the analysis, the portion of the strong solvent is increased.

5. The mobile phase needs to be prepared correctly to ensure the best possible chromatographic performance.

References

1. S. Mabic, C.Regnault, J. Krol, *LC•GC Eur.*, **18**(7), 2005, 'The Misunderstood Laboratory Solvent: Reagent Water for HPLC'.

Further reading

'Useful Solvent Characteristics', Appendix 5, Vogel's Textbook of Practical Organic Chemistry, 5th ed., Longman Scientific & Technical, 1989.

A series of articles about mobile phase buffers was featured in the magazine, LC•GC (North America version) which contains useful information regarding all aspects of using buffers in mobile phases for HPLC:

G.W. Tindall, *LC•GC North America*, **20**(11), 1028-1032, 2002, 'Mobile Phase Buffers, Part I – The Interpretation of pH in Partially Aqueous Mobile Phases'.

G.W. Tindall, *LC•GC North America*, **20**(12), 1114-1118, 2002, 'Mobile Phase Buffers, Part II – Buffer Selection and Capacity'.

G.W. Tindall, *LC•GC North America*, **21**(1), 28-32, 2003, 'Mobile Phase Buffers, Part III – Preparation of Buffers'.

Notes

The HPLC System

Instrumentation is required to enable the flow of the mobile phase through the stationary phase and also to convert the separated components into meaningful information. A typical configuration of a HPLC system is shown in Figure 12 and the main components of a HPLC system are described in Table 9.

Figure 12 Configuration of a typical HPLC System

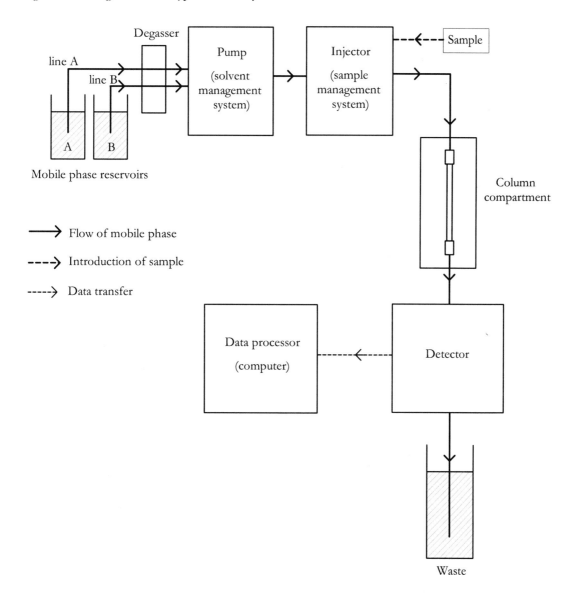

Table 9   Main components of a HPLC system

System component	Description
Mobile phase reservoir	Stores the mobile phase required for analysis
Degasser	Degasses the mobile phase
Pump	Solvent delivery system, enables the flow of the mobile phase through the system
Injector	Sample delivery system, introduces the sample to the system
Column compartment	Used to control the temperature of the column, if required
Detector	Detects each component in a separated mixture after it has eluted from the column
Data processor	Converts the data from the detector into meaningful results
Waste	Collection of the liquid waste

Each component of the HPLC system is discussed below in the order of the flow path of the mobile phase.

Mobile phase reservoir

The mobile phase is usually stored in glass containers, often these are plastic coated as a safety measure. Plastic containers are not used since additives in the plastic may leach into the mobile phase. The container needs to be of an appropriate size so that it contains enough mobile phase for the analysis being performed (e.g. 1, 2 and 5 litre flasks are often used). PTFE tubing (or a similarly inert tubing material) connects the contents of the reservoir with the HPLC system. This tubing is typically of outer diameter (OD) $^1/_8$ inch and of inner diameter (ID) $^1/_{16}$ inch. The size of the tubing in a HPLC system is usually measured using the imperial system of inches (contrasting with the column which uses metric measurements).

At the end of the tubing which is in contact with the mobile phase there is usually a filter (10 μm) to remove any particulate matter, this also acts as a 'sinker' to hold the tubing at the bottom of the container. This is commonly glass, stainless steel or PEEK. A diagram of a typical mobile phase reservoir is shown in Figure 13. A lid on the container needs to allow a space for the tubing, purpose made lids can be purchased. It is important not to seal the reservoir too tightly to avoid the creation of a vacuum.

The number of mobile phase reservoirs will depend on the number of lines available on the instrumentation. To perform reversed phase gradient elution more than one line is required so that the proportion of the organic component in the mobile phase can be increased throughout the analysis. Isocratic (1 line), binary (2 lines), ternary (3 lines) and quaternary (4 lines) systems are common. In Figure 12 a binary system is shown. The mobile phase reservoirs and the lines from these to the HPLC system are labelled A and B.

Figure 13 Mobile phase reservoir

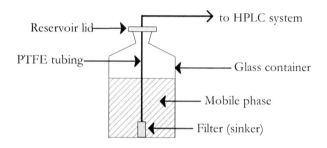

Connections

The tubing from the mobile phase reservoir is connected to the vacuum degasser if there is one installed on the system. If not, it is connected to the pump.

Degasser

Modern instruments commonly have a vacuum degasser in the flow path of the mobile phase prior to the pump. The mobile phase flows through a membrane which allows the gases to penetrate into the surrounding vacuum.

Connections

After degassing, the mobile phase is transferred from the degasser to the pump in PTFE tubing (or a similarly inert tubing material), typically $^1/_8$" OD and $^1/_{16}$" ID.

Pump

The pump is used to manage the solvents in the HPLC system. It pumps the mobile phase through the system at a controlled flow rate and at high pressures. The pump also performs online mixing of different components of the mobile phase as required. The rate of the flow of mobile phase provided by the pump needs to be accurately controlled to perform HPLC analysis. The actual flow rate depends on the scale of the HPLC system being used. Previously, in Chapter 2, the nomenclature for the different sizes of internal diameter with respect to HPLC columns was discussed. A similar nomenclature is used to describe the different scales of HPLC systems. The typical ranges of flow rates for different scales of HPLC systems are presented in Table 10.

Table 10   Typical ranges of flow rate for different scales of HPLC systems

HPLC system	Approximate typical flow rate
Analytical	0.3 to 10.0 mL/min
Micro	50 to 1000 μL/min
Capillary	0.4 to 200 μL/min
Nano	25 to 4000 nL/min

Reciprocating single piston pump

The most common type of pump for analytical scale systems is a reciprocating single piston pump. A schematic diagram to illustrate how this works is presented in Figure 15. A piston is moved in and out by the use of a motor. During the suction part of the operation (a) mobile phase is sucked through the inlet check valve at the bottom of the assembly and into the pump head, the inlet check valve is open but the outlet check valve is shut. Check valves are one way valves which allow the liquid to flow in one direction only. This is achieved by the use of a ball and seat design, these come together to prevent flow in one direction but come apart to allow flow in the other direction. During the discharge part of the operation (b) the mobile phase in the pump head is pushed out through the outlet check valve, the inlet check valve is now shut. A seal around the piston prevents leakage of the mobile phase out of the pump head.

If the motor moves the piston at a constant speed then the flow of liquid which would result from this single piston pump would be discontinuous and would pulse as shown in Figure 14. During suction the output of mobile phase would be zero. During discharge the output would start at zero build up to the maximum and then reduce to zero again.

Figure 14 Output from a reciprocating single piston pump

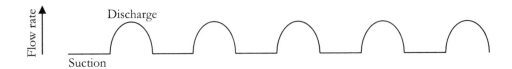

This type of flow is unsuitable for use in HPLC analysis. Pulses in flow causes problems with the detector and prevents the use of quantitative analysis, it also leads to premature failure of columns. Therefore some type of 'pulse dampening' is

required. If the piston stroke is altered to perform suction very quickly and discharge slowly then some of the pulsation may be reduced.

Figure 15 Schematic of the operation of a reciprocating single piston pump

(a) Suction

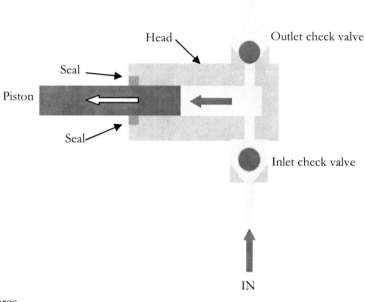

IN

(b) Discharge

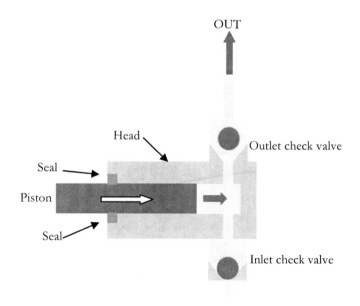

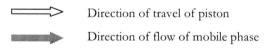

Direction of travel of piston

Direction of flow of mobile phase

Pulsed flow may also be reduced by the use of two piston pumps. The two pistons are 180 degrees out of phase and thus one piston is delivering mobile phase while the other is filling and vice versa. Tandem designs where the pump pistons are in series may also be used. There are many sophisticated design solutions to provide a uniform flow and most of the differences in pumps from different manufacturers are related to design of the pistons and other pulse dampening solutions.

Particulate matter and air in the mobile phase can cause problems with the check valves and thus for trouble free operation it is best to filter and degas mobile phase. Also buffer salts can crystallise out causing damage to the check valves and the pump seals. To avoid this is it a good idea to flush systems thoroughly with buffer free mobile phase after using buffers, especially non-volatile buffers. Most repairs required for the pump will be related to the check valves and the pump seals. These should be replaced periodically and for this reason pumps are designed to make them easily accessible.

Different pumping designs may be used for sub-micro scale pumps because the flow rates required are very small. One design is a syringe pump where the mobile phase is delivered by a syringe action. Different manufacturers have different solutions for pumping mobile phase at low flow rates.

Back pressure

A high pressure is experienced in the HPLC system which results from forcing the mobile phase through the densely packed column. This is referred to as the **back pressure** (the tubing in the HPLC also contributes slightly to the back pressure). The pump needs to be able to perform at high pressures to deliver a constant flow rate, regardless of the back pressure. During gradient separations the back pressure will usually change over the course of the analysis due to the alteration of the viscosity of the mobile phase, the flow rate needs to be constant.

The magnitude of the pressure will depend upon the rate at which the mobile phase is pumped and also the physical characteristics of the column (i.e. the particle size, the internal diameter and the length of the column). The units used to measure pressure vary between different manufacturers of HPLC systems. They are summarised in Table 11. A useful rule of thumb is 1000 psi ≈ 70 bar, 100 bar = 1450 psi

Table 11   Pressure units used in HPLC

Pressure units	Name of units	1 unit equivalent to:			
		MPa	bar	psi	Atm
MPa	Megapascal	-	10 bar	145 psi	9.87 atm
bar	Bar	0.1 MPa	-	14.5 psi	0.987 atm
psi	Pounds per square inch	0.007 MPa	0.069 bar	-	0.068 atm
atm	Physical atmosphere	0.101 MPa	1.013 bar	14.7 psi	-

Recent advances in technology have resulted in particle sizes of less than 2 microns being used in HPLC columns. To pump mobile phase through these columns at an optimum flow rate requires the use of very high pressures. Pump designs have been modified to enable this and instrumentation is commercially available to operate at very high pressures. Typically standard pressure HPLC systems operate up to approximately 40 MPa (megapascal, the SI unit for pressure) and very high pressure systems operate up to approximately 100 MPa.

Online mixing

Gradient elution requires that the composition of the mobile phase is altered during an analysis therefore the pump needs to be able to mix solvents. This may be achieved using low pressure mixing, where the solvents are mixed prior to the pump, or by high pressure mixing, where they are mixed after the pump and thus more than one pump is required. High pressure mixing systems offer slightly better control and are often used with very sensitive detectors. Two pump systems are commonly referred to as 'binary pumps' by manufacturers. Low pressure mixing systems offer a more flexible choice of solvents since up to four solvents can be mixed using one pump and a proportioning system. Low pressure systems are more common although many manufacturers offer both options.

The time taken for a change made in the composition of the mobile phase to arrive on the column means that there is a slight delay between changing the gradient and its effect being observed. This delay is due to the **dwell volume** of the instrument. The dwell volume for low pressure instruments is higher than that for high pressure instruments since there is further for the mobile phase to travel. The dwell volume is important to consider during gradient analysis since there will be a delay between changing conditions and seeing the effect.

Connections

When the mobile phase emerges from the pump it is under pressure. The tubing used is usually made of stainless steel but PEEK may also be used. The diameter of the tubing is typically $^1/_{16}$" OD and 0.010" or less ID (also commonly referred to as '10 thou', an abbreviation of '10 thousandths of an inch'). This tubing is connected to the injector.

Injector

The function of the injector is to introduce the sample into the mobile phase so that it can be separated on the HPLC column. The injector needs to be able to accurately introduce the desired volume of the sample into a pressurised system. The amount of sample which can be injected is related to how much packing material is in the column being used. Typical injection volumes for the different scales of HPLC are given in Table 12. Injection volume is usually measured in microlitres (µL).

Table 12   Typical injection volumes for HPLC

HPLC system	Typical injection volume (µL)
Analytical	20
Micro	5
Capillary	0.1
Nano	0.01

6 Port injection valve

Most injectors use a 6-port injection valve, as shown in Figure 16, to introduce the sample without stopping the flow of mobile phase. When the valve is in the LOAD position, as shown, the mobile phase is flowing through the valve from position 4 to position 5 and then to the column. The syringe containing the sample is placed in position 1 (a needle port) and the sample is injected into the sample loop (attached between positions 6 and 3). The sample loop contents are completely replaced by the sample and the waste is removed via position 2. The valve is then moved into the INJECT position. Now the mobile phase flows through the sample loop and delivers the sample to the column. The needle port, position 1, is now connected to the waste, position 2, and can be rinsed before returning to the LOAD position for the next injection.

Figure 16   6-Port injection valve

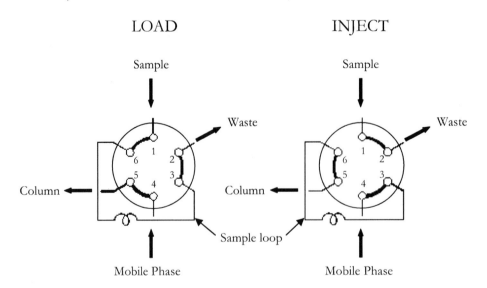

Sample loops are made in such a way so as to contain a precise volume, if the sample loop in Figure 16 was of volume 20μL, then exactly 20μL would be injected. This is called 'filled loop' injection. If less than 20μL is injected into this sample loop it can still be delivered to the column, however the amount would need to be accurately introduced to the sample loop. This is known as 'partial loop' injection and is commonly used by autoinjectors to enable the analysts to select the injection volume that they require without changing the sample loop.

If the analyst is manually injecting a sample by injecting with a HPLC syringe into a 6 port valve then the filled loop injection will deliver a more accurate injection volume. (Note: HPLC syringes have a square tip and are designed for used with injection valves.) If the accuracy of the injection volume is not critical then partial loop injections can be performed by injecting the desired amount using the syringe (as opposed to flushing through completely with filled loop injection). This may be more convenient than changing the sample loop.

Autoinjectors

When an analysis involves a large number of injections manual injection is not viable. This is commonly the case when quantitative analysis is being performed on samples during drug development, or quality assurance of the manufactured product. An automatic injection system is required. Autoinjectors, also known as autosamplers, withdraw the sample from a capped vial and inject it into the mobile phase following a pre-programmed sequence which contains the details of the injection volume, the vial number and the time between injections.

Designs of autoinjectors vary between manufacturers and, as for pumps, they are optimised for different scales of HPLC. Most are based on the principle of the 6-port valve described above. The flexibility of programming features and the number of available sample positions will vary between different manufacturers.

The sample is usually injected from a vial. These can be a variety of sizes but the most common are 2mL vials (see Figure 17a). They can be sealed using snap on, crimped or screw top caps. The needle of the autoinjector pierces the cap during the injection process and withdraws the required aliquot from the vial. Other sizes of vial are available when using most brands of autoinjector, and inserts can be used when the amount of sample is limited (see Figure 17b). Many autoinjectors can also accept well plates and injections are performed directly from these.

Figure 17   Autoinjector vials

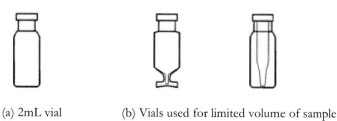

(a) 2mL vial   (b) Vials used for limited volume of sample

Most autoinjectors have some type of wash system. The design of the injection system will determine which parts of the instrument are in contact with the sample and therefore need to be rinsed. Usually a wash solvent which is known to dissolve the sample of interest is used. Complicated wash programmes are available on some instruments, which can be useful for difficult sample matrices.

Connections

The mobile phase, carrying the sample, emerges from the injector in stainless steel (or PEEK) tubing of diameter $^1/_{16}$" OD and 0.010" or less ID. There may be an inline filter installed on the tubing to protect the column from pump-seal and valve-seal particles as well as those originating in the mobile phase or sample. This filter consists of a stainless steel frit, typically of pore size 2 μm. The tubing from the inline filter is then connected directly to the guard column, if in use, or the HPLC column.

Column compartment

The column compartment is the part of the HPLC system where the column is housed. Most modern instruments include a column compartment so that the temperature can be controlled during HPLC analysis.

Column fittings

The column (and guard column if in use) is attached to the tubing from the injector using fittings as shown in Figure 18. These may be made from stainless steel or from polymeric materials such as PEEK.

Figure 18 Examples of fittings for connecting tubing to the HPLC column

(a) Stainless steel male nut and ferrule   (b) One piece PEEK fitting   (c) PEEK male nut and ferrule

The stainless steel fitting (a) is comprised of two parts: the 'male' nut which is threaded and matches up with the threading on the 'female' end fitting of the column, and the ferrule which ensures the integrity of the connection. The nut may be tightened with a wrench or is also available as a finger-tight fitting (i.e. are tightened by hand without the use of a wrench). The first time that the ferrule is used on the tubing will determine the position of the ferrule in the future. When the fitting is tightened in the end fitting of a column the ferrule will be irreversibly attached to the tubing and the seating depth as shown in Figure 19 will be set. This seating depth will depend on the column used and may vary between some column manufacturers leading to problems when changing over from one column to another. The fitting may leak, or worse, it may appear to work satisfactorily but in fact is contributing to band broadening and decreases the quality of the separation. Stainless steel fittings should not be used on PEEK tubing.

Figure 19 The seating depth of a stainless steel ferrule

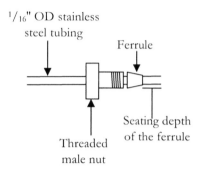

Polymeric fittings do not irreversibly attach to the tubing and thus the seating depth will adjust to the column in use. They are available in one-piece and two-piece configurations as shown in Figure 18. Polymeric fittings are also available in wrench tightened or finger-tight options. The choice of which type of fitting to use is usually based on personal preference but also compatibility with the temperature and pressure being used must be considered. For some biological systems stainless steel cannot be used and polymeric fittings are preferable.

Care must be taken not to over-tighten either stainless steel or polymeric fittings since this can damage the threading on the fitting and on the column end-fitting making it difficult to obtain a good connection the next time they are used. Similar fittings are used where connections of the tubing are required throughout the HPLC system in the pump, injector and detector.

If a guard column is in use it is connected before the column using the fittings described. If a cartridge system is being used then this will connect directly to the column. Otherwise a column coupler is used. This consists of two fittings, as described previously, attached to a short length of $^1/_{16}$" OD tubing.

Temperature control

The column compartment is usually temperature controlled because temperature is an important variable when performing HPLC analysis. The inlet tubing to the column is also heated so that the mobile phase enters the column at the desired temperature. Temperatures of between 30 and 50°C are common although recent advances in column technology may mean that the use of high temperature HPLC (up to 200 °C) becomes popular in the future.

Connections

The outlet tubing from the column goes to the detector in stainless steel (or PEEK) tubing of diameter $^1/_{16}$" OD and 0.010" or less ID.

Detector

The detector is placed at the outlet of the column and the mobile phase carrying the separated analytes is passed through it. The detector needs to able to detect when an analyte is present. The time taken for each of the sample components to elute is measured and information is also obtained about the amount of each component which is present.

The output from the detector is converted into an electric signal in order to display it. The signal is constant as the mobile phase flows through it. When an analyte is detected the signal will rise to a maximum level and then decrease back down to the constant level obtained for the mobile phase. For an ideal separation the analyte molecules will be in a normal distribution therefore the shape of the signal from the detector will be Gaussian as shown in Figure 20.

Figure 20 The expected detector signal for an ideal separation

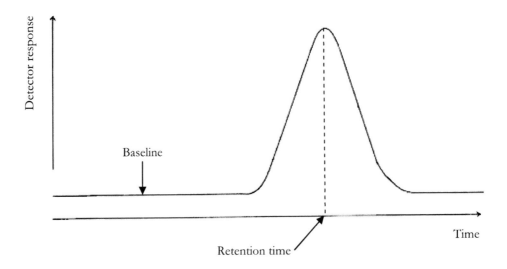

In HPLC this is referred to as a **peak**. The constant level due to the mobile phase is referred to as the **baseline**. The magnitude of the signal at each time point across the peak is related to the amount of analyte present. The time at the maximum of the peak is known as the retention time of the analyte.

Some of the desired characteristics for a HPLC detector are:

1. Must be able to detect the presence of analytes in the mobile phase. This is achieved by either detecting a change in the bulk property of the solution in the detector (non-selective) or by detecting a property of the analyte (selective). Detectors that can obtain a consistent relationship between the magnitude of response and quantity injected for a range of different analytes are referred to as 'universal' detectors. A universal detector will detect all the components in a mixture.

2. A linear relationship where the magnitude of the signal is proportional to the concentration of the analyte is required for quantitative analysis. An ideal detector will have a wide range of linearity.

3. Able to detect small amounts of analyte.

4. Not affected by changes in temperature and mobile phase composition.

5. Not contribute to band broadening, therefore the area of the detector where the measurement takes place should be as small as possible.

6. Able to detect narrow peaks generated during fast analyses correctly.

There are a variety of detectors that can be used with HPLC. Examples of some available detectors are given in Table 13. The most common HPLC detector used for pharmaceutical analysis is the UV detector.

Table 13   Examples of HPLC detectors

Detector	Description
Ultra Violet (UV)[1]	Detection is based on absorption of UV light by an analyte and therefore is applicable to analytes which have this property. Different compounds do not absorb do the same extent or at the same wavelength and therefore this is a selective detector. It is a robust, sensitive detector which is suitable for gradient analysis and has a wide linear range.
Diode array	A UV detection technique where an array of diodes are used to monitor the full UV range simultaneously. Thus it provides a UV spectrum at any time point during the analysis.
Mass Spectrometer (MS)	A mass spectrometer separates ions in a vacuum according to their mass-to-charge ratio, it uses electrical or magnetic fields, or a combination of both, to move the ions from the region where they are produced, to a detector, where they produce a signal which is amplified. Molecular weight data on analytes can be obtained. MS is not just a HPLC detector, the hyphenation of the two techniques, referred to as LC-MS, provides a very powerful analytical tool for pharmaceutical analysis.

Detector	Description
Refractive Index	Detection is based on changes of refractive index when the analyte passes through the sample cell in the detector, the reference detector being filled with the mobile phase. It is a universal detector but is very sensitive to mobile phase composition and temperature. This makes it unsuitable for gradient analysis.
Evaporative Light Scattering (ELS)	Detection is based on the scattering of a beam of light by particles of analyte remaining after evaporation of the mobile phase. It is a universal detector and can be used with gradient elution. The use of volatile buffers is required since the mobile phase is evaporated.
Electrochemical	The electrochemical detector responds to substances that are either oxidisable or reducible and the electrical output is an electron flow generated by a reaction that takes place at the surface of the electrodes. It can be applied when the analyte of interest possesses an electroactive polar constituent such as a hydroxyl, amine, or sulfur-containing group. Electrochemical detection is a very sensitive detection technique but can be difficult to operate successfully.
Fluorescence	Detection is based on fluorescent emission following excitation of a fluorescent compound at an appropriate wavelength. It is only applicable to compounds exhibiting fluorescence and derivatives of these but is very sensitive.
Corona® CAD® (charged aerosol detection)	Based on established evaporative techniques that have been deployed effectively in ELS detectors and LC-MS systems. However, the charged aerosol detector is a new detection methodology for HPLC. A universal detector and suitable for use in gradient analysis.
Conductivity	The electrical conductivity detector measures the conductivity of the mobile phase, it can only detect those substances that ionise and therefore may be used to detect charged species in ion exchange chromatography.

Ultra Violet (UV) Detectors

The presence of an analyte in the mobile phase may be detected if it absorbs UV light. Functional groups within molecules which can absorb light in the UV and visible regions of the electromagnetic spectrum are known as **chromophores**. These functional groups contain electrons which are shared in double or triple bonds and also those which are localised about such atoms as oxygen, sulfur, nitrogen and the halogens. These electrons can be excited in the region 190 to 800 nm providing useful absorption data. The electrons in single bonds require more energy and thus lower wavelengths (< 180 nm) to achieve excitation. The wavelength at which the absorption is at a maximum is referred to as the lambda max, λ_{max}. This will vary for different functional groups. The magnitude of the absorption also varies for different functional groups. It depends on the molar absorptivity, ε, for the molecule.

Examples of chromophores with their values for λ_{max} and ε are presented in Table 14. This list is intended as a guide, not an exhaustive list, the value for λ_{max} will vary with the composition and pH of the mobile phase used for analysis. A comparison of the absorption expected for different functional groups can be obtained from the molar absorptivity values. The consequence of the difference is that if the detector detects two peaks of equal size for a separation involving two components this does not necessarily mean that there is the same amount of each in the mixture. **Relative response factors** may be calculated by comparing the detector response obtained for known concentrations of each component.

Table 14   Examples of common chromophores

Chromophore	λ_{max} (nm)	ε
-C=C-	190	8,000
C=C-C=C	219	6,500
-(C=C)$_3$	260	35,000
C=C-C=N	220	23,000
-NH$_2$	195	2,800
-N=N-	285 - 400	3 - 25
-COOH	200 - 210	50 - 70
-COOR	205	50
>C=O	195	1,000
>S-O	210	1,500
-N=O	302	100
	184	46,700
	202	6,900
	255	170
	220	112,000
	275	175
	312	5,600

The relationship between magnitude of response for UV detection and the concentration of the sample is defined by Beer's law:

$$A = \varepsilon \, l \, c$$

Where 'A' is the absorption, 'ε' is the molar absorptivity, 'l' is the pathlength and 'c' is the concentration. The absorption is therefore directly proportional to the concentration of the analyte if 'l' is kept constant. The pathlength is defined by the flow-cell in the detector where the absorption takes place. Beer's law applies at absorption less than approximately 2 absorption units (AU), this provides a large range of linear response which can be used for quantitative analysis.

Modern UV detectors are variable wavelength detectors, this means that any wavelength in the UV range can be selected for detection. A deuterium lamp is used to provide the UV light up to about 340 nm and a tungsten lamp may be used to extend into the visible region up to approximately 800 nm. Some detectors allow multiple wavelengths to be collected at the same time. The lamp will need to be replaced routinely and the instrument is usually designed to make it easily accessible. The flow-cell may also be available in different sizes for different scales of HPLC and in this case will also be accessible.

Diode array detectors also detect based on the absorption of UV light but the design of the instrument is different. A photodiode array is used which collects absorption data for all wavelengths in the range of approximately 190 to 800 nm. This means that a UV/Vis spectrum can be obtained for any time during an analysis. This information is useful for identification and characterisation of sample components. It is also very useful during HPLC method development when assessing the best wavelength to use for a separation.

Mobile phase requirements for UV detection
Since the response for a UV detector is based on the absorption of UV light it is important that the mobile phase does not also absorb UV light. The UV cutoff is the wavelength above which a solvent is not UV active. The UV cutoff values for common solvents and buffers are given in Table 15 and Table 16 respectively.

Table 15   UV cutoffs for common solvents

Solvent	UV Cutoff (nm)	Solvent	UV Cutoff (nm)
n-hexane	200	Isopropanol	210
Methylene chloride	233	Acetonitrile	190
Chloroform	245	Methanol	210
Methyl-t-butyl ether	210	Water	190
Tetrahydrofuran	215		

Table 16   UV cutoffs for common buffers

Buffer	UV Cutoff (nm)	Buffer	UV Cutoff (nm)
Trifluoroacetic acid (TFA)	210 (0.1%)	Carbonate, pK_1	< 200
Phosphate, pK_1	< 200	Phosphate, pK_2	< 200
Citrate, pK_1	230	Tris-(hydroxymethyl)aminomethane	205
Formate	210 (10 mM)	Ammonia	200 (10 mM)
Citrate, pK_2	230	Carbonate, pK_2	< 200
Acetate	210 (10 mM)	Triethylamine	< 200
Citrate, pK_3	230	Phosphate, pK_3	< 200

Solvents and buffer may be used for analysis at wavelengths above the UV cutoff value.

Mass Spectrometer (MS)

Mass spectrometry is a very powerful analytical technique in its own right rather than just a detector for HPLC. HPLC may be considered a sample preparation technique for MS. The hyphenation of these two techniques, referred to as LC-MS, enables the collection of molecular weight information on the separated components in a mixture. A full discussion of MS is outside the scope of this book, it is a complex analytical tool[2]. What follows is a brief overview of LC-MS[3].

The outlet from a HPLC system is a flow of mobile phase containing separated analytical components. The mass spectrometer analyses gas phase ions in a vacuum. Therefore to connect HPLC and MS involves the transformation of a solute into a gas phase ion whilst getting rid of the solvent and maintaining adequate vacuum level in the mass spectrometer. Interfaces for connecting HPLC and MS have been developed to solve this difficult problem. The interfaces which are most commonly used are Electrospray (ES) and Atmospheric Pressure Chemical Ionisation (APCI). Both interfaces operate at atmospheric pressure.

In electrospray a charged aerosol is generated and a flow of nitrogen is used to strip away the solvent. The charged molecules are moved using electrostatically charged plates and are directed into the MS. In APCI a corona discharge is used to produce ions such as H_3O^+ and N_2^+, which promote ionisation of the sample. The two techniques tend to be complementary since electrospray is more suitable for polar molecules and APCI will ionise less polar molecules. MS instruments that are designed to be used with HPLC are designed in such a way that either of these ionisation techniques may be used. Electrospray and APCI may be used with flow rates of up to 1 and 2 mL/min respectively. However electrospray performs best at a flow rate of 200 μL/min or below. A flow splitter may be used for higher flow rates to divert a suitable portion of the HPLC outlet to the MS. The remainder is sent to waste or may be used with another detector.

In pharmaceutical analysis LC-MS is used in two main ways although these may overlap to some extent.

1. It is used qualitatively to obtain molecular weight and structural information on drug related molecules. This information is critical during the drug discovery and development processes and thus LC-MS is used extensively for this purpose.

2. It is used quantitatively as a sensitive and selective detector for a specific target molecule. LC-MS is not normally used for the type of routine quantitative analysis typical of a quality assurance laboratory. However it may be used in specific methods where trace amounts of an analyte need to be detected, e.g. environmental analysis of drugs in effluent.

Mobile phase requirements for MS detection
Due to the nature of the LC-MS interface it is important that the mobile phase is volatile. Inorganic buffers precipitate out of solution and may cause damage to the instrument. Volatile buffers which are suitable for use with LC-MS include: trifluoroacetic acid, formate, acetate, carbonate, tris, ammonia and triethylamine.

Connections

The outlet from the detector is on the low pressure side of the HPLC system. Small gas bubbles can form as solvent moves from the high pressure to the low pressure environment, causing problems with the detector. For this reason a short length of tubing of $^1/_{16}$" OD and 0.010" or less ID is often used to create a back pressure in the detector. A device called a 'back pressure regulator' may also be used for this purpose. $^1/_8$" OD PTFE type tubing is then connected to the $^1/_{16}$" OD tubing or the back pressure regulator and is used to carry the waste to a suitable receptacle.

Data Processor

The peaks generated by the detector need to be manipulated to obtain useful analytical results and also to obtain a record of the analysis. The electronic signal from the detector is sent to a data processor. This is typically a computer which has suitable software to allow processing of the data. The graphical representation of an analysis is called a **chromatogram**. This displays the output of the detector plotted against time for one injection. An example of a chromatogram is shown in Figure 21.

Figure 21 Example of a chromatogram

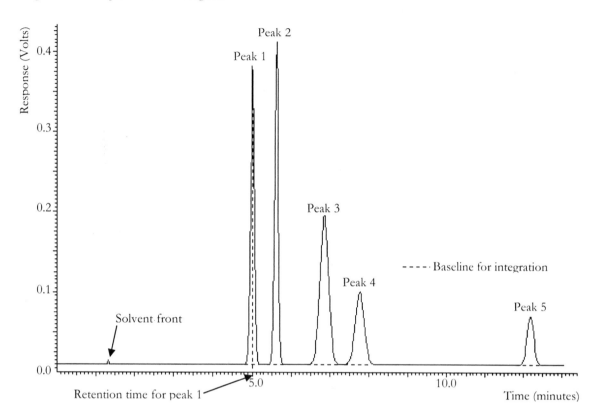

The features of the chromatogram in Figure 21 are:

1. The peaks present in the chromatogram correspond to analyte passing through the detector. In the example there are 5 well separated peaks. The width of the peak increases to give broader peaks with increasing time due to diffusion effects in the column.

2. The time at which a peak is detected is called the retention time. This is measured at the apex of the peak as shown for peak 1 in Figure 21. The retention time for peak 1 is 5.0 minutes.

3. A small peak is shown in the example at 1.3 minutes due to the solvent front. This is the detector response for unretained material in the sample. The time taken for unretained material to travel through the column is referred to as t_0 (in the example $t_0 = 1.3$ minutes), it is caused by the **void volume** of the column. The solvent front may also be a small disturbance in the baseline, a large peak or may be negative, it will depend on the solvent used for the sample and the detector settings.

4. The axes for a chromatogram are time (usually measured in minutes) along the horizontal axis and detector response along the vertical axis. The detector response may be in volts since it is converted into an electronic signal or may be

displayed as the equivalent in the units of measurement for the detector, e.g. the vertical axis in Figure 21 is measured in units of volts. If the detection method was UV and the data processor was set to convert 1 A.U. into 1 volt then the label on this axis could alternatively read 'absorbance (A.U.)'. The user can usually define the desired detector settings.

5. To perform quantitative analysis it is necessary to measure the size of the peak so that it can be compared with a standard of known concentration. The area and the height of the peak may be used for this purpose but use of the area is much more common. In order to use the area the peak must be integrated, i.e. the area under the peak must be measured. This mathematical process is easily accomplished using computer software but the analyst needs to define the area to measure. The baseline of the chromatogram is the detector response where only mobile phase is present. To define an area for integration it is necessary to continue this baseline under the peaks as shown in Figure 21.

Chromatography Data Systems (CDS) are computer software packages that provide a user interface to control the instrumentation, process the analytical data and manage the results for HPLC analysis. CDS are available from manufacturers of HPLC instruments and also from software specialists. Important features of CDS include:

Instrument control & data acquisition
The operating parameters of the HPLC instrumentation may be controlled remotely from a computer. Some CDS give full control of the instrumentation whilst others offer limited control only. In recent years manufacturers have collaborated to allow control of their instruments by numerous CDS packages. This is especially useful in laboratory where a variety of brands of instrumentation are used. The sequence of injections required for an analysis is set up on the CDS and the data obtained for each injection is then stored using the parameters input by the user.

Processing data
CDS are used to manipulate chromatograms in order to obtain useful data. They facilitate naming peaks, integration of peaks[4] and the production of quantitative data by use of calibration functions. Chromatograms can be overlaid and subtracted from each other and a range of statistics may be calculated. The data associated with MS detection involves complex 3D data, therefore software is available which is specifically for LC-MS. The 3D data associated with photodiode array detection can be manipulated using some CDS systems.

Management of results
The data generated by a HPLC system can be vast and therefore a suitable management system is required which can be easily used to find a particular result, sample, chromatogram etc. Most CDS will enable the conversion of results into user defined reports and also the integration of data with other software packages such as Microsoft Excel®.

Regulatory compliance

The pharmaceutical industry is a highly regulated industry and electronic data needs to comply with the FDA regulations '21 CFR part 11'. This defines how electronic records should be managed and it is important that a CDS enables the functionality required for compliance.

Networking

In large laboratories networking of instruments is desirable for ease of access to data and also system administration. Most CDS packages are designed so that they can be used standalone (a single computer workstation with one or more instruments attached) or in a network. Some CDS can be linked to Laboratory Information Management Systems (LIMS).

Refer to Table 17 for some examples of CDS packages. The difference between these packages relate to how data is managed and displayed, instrumentation that is controlled and additional functionality such as method validation.

Table 17 Examples of chromatography data systems

Manufacturer	CDS
Agilent	Chemstation Plus®, Cerity®
DataApex	Clarity™
Dionex	Chromeleon®
Perkin Elmer	TotalChrom™
SSI	EZChrom™
Thermo	ChromQuest™
Varian	Galaxie™
Waters	Empower™, MassLynx™

System hardware

There are many manufacturers of HPLC systems. Usually the entire system is available for purchase from one supplier and it is more convenient to obtain a system in this way rather than using different suppliers for each component. However, some add-on components such as specialised detectors may be sourced separately. Systems are usually constructed in a modular format so the components described in this chapter are readily identifiable. Different modules may be selected depending on the requirements for the system.

The components in a HPLC system are often optimised for a particular use, e.g. micro, capillary and nano HPLC require instruments which can deliver low rates and have low system volume. This is particularly important in the case of very small scale columns where the column volumes are very small. To get the full benefit of these columns the extra-column volume needs to be minimised and systems are designed with this in mind. Systems are also designed specifically for use with size-exclusion chromatography and ion-exchange chromatography and these are equipped with the most suitable detectors, etc. for these types of HPLC. In some bioanalysis applications, proteins absorb on stainless steel and the metal ions alter chromatographic profiles and retentions. Therefore biocompatible systems where the stainless steel parts of the system are replaced with PEEK are also available from some manufacturers. Examples of commercially available systems are shown in Table 18. These systems are all available in a range of configurations.

Table 18   Examples of commercially available HPLC systems

Manufacturer	System
Agilent	Agilent 1200 series
Hitachi	LaChrom Elite®
Jasco	LC2000 Plus
LC Packings (Dionex)	Ultimate™ 3000
Perkin Elmer	Series 200
Shimadzu	Prominence™
Thermo	Accela™
Varian	ProStar series
Waters	Acquity UPLC™, Alliance

The current trend in HPLC is towards faster analyses[5], this has been achieved by advances in column technology including the use of smaller particle sizes (and very high pressure instrumentation) and also by use of shorter columns and higher temperatures.

Summary

1. The HPLC system is comprised of the following: mobile phase reservoir; degasser; pump; injector; column compartment; detector; data processor and waste vessel.

2. The pump is the solvent delivery system, it must be able to operate at high pressures, deliver mobile phase at a controlled flow rate and mix solvents accurately for gradient analysis.

3. The injector is the sample delivery system, it injects accurate amounts of sample into the flow path of the mobile phase.

4. The column is connected to the system using screw tight end fittings, it is often housed in a column compartment which is held at a controlled temperature.

5. The detector needs to be able to detect the presence of the analyte when it elutes from the column. The most common detectors in pharmaceutical analysis are UV and MS.

6. The data processor converts the data from the detector into meaningful results.

References

1. B.E. Lendi, V.M. Meyer, *LC•GC Eur.*, **18**(3), 2005, '**The UV Detector for HPLC – An Ongoing Success Story**'.

2. J.H. Gross, '**Mass Spectrometry – a Textbook**', Springer, 2004.

3. M. McMaster, '**LC/MS: A Practical User's Guide**', Wiley, 2005.

4. N. Dyson, '**Chromatographic Integration Methods**', Royal Society of Chemistry, 1990.

5. C. Mann, *LC•GC Eur.*, **20**, 290-299, 2007, '**Ultrafast HPLC: Different approaches to increased throughput**'.

Further reading

The websites of HPLC system manufacturers contain useful information about their products, these may be found easily using internet search engines.

Notes

Chapter

5

The HPLC Analytical Method

I n chapters 1 to 4 the system required to perform HPLC analysis comprising the stationary phase (the column), the mobile phase and the instrumentation was described. To define how all these elements will be used for a particular application requires the use of a HPLC analytical method. This contains all the necessary information that an analyst will need to perform the analysis.

Applications of HPLC in pharmaceutical analysis

To understand the purpose of the HPLC analytical method it is necessary to consider the applications of HPLC in pharmaceutical analysis. There are a wide variety of applications throughout the process of creating a new drug, from initial drug discovery to the manufacture of formulated products which will be administered to patients. This process[1,2] is summarised below and the applications of HPLC are discussed at each stage.

The creation of a new drug

The process to create a new drug can be divided into three main stages. These are drug discovery, drug development and manufacturing.

Drug discovery

The first step in the drug discovery stage is to select the disease which the drug will treat. This decision will be based on unmet medical needs and also on financial considerations since a huge investment is needed to bring a new drug to market (this has been estimated at almost £500,000).

Drug target

The next step is to identify a suitable **drug target**. Drug targets are areas in the cells of the body where drugs attach and result in a therapeutic effect. The major drug targets in the body are normally large molecules (macromolecules) such as proteins, nucleic acids, lipids and carbohydrates. An understanding of which of these is involved in a particular disease state is an important starting point for a new drug project. The human genome project which has mapped the DNA of humans, and also proteomics which studies the role of proteins and their effects on diseases, are revealing an ever increasing number of new proteins which may act as potential drug targets. Typically a number of targets will be tested in a new drug project.

When a potential drug target has been identified the challenge is to find a chemical which will interact with the target. To evaluate the effect of chemicals on the target a **bioassay** is required. This is a test to check if the desired interaction is taking place. **In vitro** tests, where specific tissues, cells or enzymes are used, are performed which are designed to produce an easily measurable effect when interaction occurs, such as cell growth or an enzyme catalysed reaction which produces a colour change.

High throughput screening (HTS) involves the automated testing of large numbers of compounds versus a large number of targets, typically several thousand compounds can be tested at once in 30 - 50 biochemical tests. Assays are run in a parallel fashion using multi-well assay plates (e.g. 96-, 384- and 1536-well). While the actual screen may only take a few days the design of the bioassay and automation may take much longer. The compounds screened in HTS may be from an existing compound 'library', a collection of compounds which have been synthesised by a pharmaceutical company over many years of research. They may also be sourced from **combinatorial synthesis**, a method of producing a large number of compounds in a short period of time, using a defined reaction route and a large variety of starting materials and reagents. It is usually performed on a very small scale to allow automation of the process.

The result of screening is the identification of 'hits', compounds active in the screens that have the potential to be made into drugs. These hits (numbering in the hundreds) are further analysed and screened to reduce their number, the result being lead compounds. A **lead compound** is a structure which shows a useful pharmacological activity and can act as the starting point for drug design.

Applications

LC-MS is a useful tool for compound identification and characterisation. It may be used as a measurement tool during high throughput screening. Preparative HPLC is also used to isolate and purify hits and lead compounds as required, e.g. from combinatorial synthesis.

Drug design

Once the structure of a lead compound is known its **structure-activity relationships** (SARs) are studied using biological testing. This defines the functional groups or regions of the lead compound which are important to its biological activity. At this stage a range of compounds which are similar to the lead compound are synthesised and tested to optimise the drug target interaction. The study of how the drug interacts with the target is called **pharmacodynamics**.

The drug not only has to interact with the target, it also has to reach the target in the body once it is administered. This area of study is known as **pharmacokinetics**. Structural modifications to the lead compound are investigated to improve its ability to reach the drug target. These studies involve investigation of absorption, distribution, metabolism and excretion of the compound and are thus often referred to as ADME studies. A large number of samples are generated from in vitro and in vivo ADME screens.

The optimisation of the pharmacodynamics and pharmacokinetic properties of the lead compound are referred to as lead optimisation. The outcome may be a number of structures which are potential candidate drugs. This series of structures will often be patented at this stage to protect the investment of the pharmaceutical company. Now a decision has to be made about which compound to take into the expensive stage of drug development.

Applications

During lead optimisation LC-MS is used for characterisation and identification of the compounds which are synthesised. LC-MS is routinely used to support ADME screens by the analysis of metabolites during pharmacokinetic testing.

Drug development

The various parts of drug development usually run concurrently since any one can take a considerable amount of time. The drug development stage is also considerably more expensive than drug discovery.

The first stage in drug development is the **preclinical trials**, the purpose of which is to study the toxicity, drug metabolism, pharmacology and a suitable formulation for the potential drug. Toxicity, drug metabolism and pharmacology testing is performed through in vitro and in vivo laboratory animal testing. Although the pharmacology of the drug may have been studied in the drug discovery stage, further study is carried out to gain a better insight into the drug's mechanism of action. Preformulation studies involve characterisation of the drug's physical, chemical and mechanical properties in order to choose what other ingredients should be used in the formulation.

Drug development

Applications

LC-MS is used for the analytical testing of the biological samples produced from drug metabolism studies. If radio-labelled drugs are used for tracing metabolites they may be detected using HPLC with a radioactivity detector. HPLC may be used to isolate and purify the material used in preclinical studies, and to obtain purity profiles of the material prior to use.

Clinical Trials

If the drug has the desired effect in animal tests and has performed well in the other preclinical trials then it may be brought into the next phase, **clinical trials**. This involves testing the drug on volunteers and patients and is split into four different phases:

Phase I – Carried out on a small number (in the range of 20 - 80) of healthy volunteers to provide a preliminary evaluation of the drug's safety, pharmacokinetics and dose levels but not intended to demonstrate whether the drug is effective.

Phase II – Carried out on patients (about 100 to 300) to establish whether the drug has the therapeutic property claimed and also to further study safety, pharmacokinetics and best dosing regimes.

Phase III – Carried out on a larger number of patients (anywhere from about 500 to 5000) to gain more information about the drug and its effects.

Phase IV – Carried out after the drug has been approved and is being prescribed to patient to treat their disease. The drug is continually monitored for effectiveness and side effects, it may be withdrawn if adverse effects are discovered.

The results obtained at each phase will determine whether the drug is taken into the next phase.

Applications

HPLC analysis is used to support the process of clinical trials throughout all four phases. It is used to test the quality of the material used in clinical trials by providing assay and impurity information and may be used for analysis of biological samples generated during the trials.

Development of drug substance

At the same time as the clinical trials are being performed the processes to manufacture the drug need to be designed. During development of the process to manufacture the drug, large amounts of material are required to support the clinical

---

trials. The synthetic route to make the drug will be designed so that it can easily be scaled up to produce large quantities.

This process starts at the laboratory bench in kilogram type quantities and then is scaled up to pilot plant scale of about 100kg before going into full scale production if the drug is approved. The requirements for the process are that it is straightforward, safe, cheap, efficient and high yielding, uses a minimum number of synthetic steps, and will consistently provide a high quality product. The stability of the material will have to be assessed to ensure that it can be stored.

Applications

HPLC analysis is very important during the development of the drug manufacture process. It is used at every synthetic step to provide information on when reactions are complete, on reaction yields, and assay and impurity determination of the material produced. LC-MS may be used during the development of the synthetic steps to identify the products of the reaction and also synthetic impurities. HPLC may also be used to analyse the starting materials used in the process.

Stability studies which are performed on intermediates in the synthetic process and also on the drug substance rely heavily on HPLC to determine which impurities are forming during storage, and to quantify the amounts of impurities present.

Development of formulated product

A method to deliver the drug so that it can find the drug target is required and for this a formulated product is designed. Following on from the preformulation work, factors such as particle size, polymorphism, pH and solubility are studied since these can influence the bioavailability of the drug.

A suitable **drug delivery** method is selected and the required formulated product is developed. The process to manufacture the formulated product must ensure that the correct amount of drug is present in each dosage unit. Compatibility studies with the excipients and the packaging are required to ensure there are no unwanted interactions with the drug. The stability of the formulated product needs to be evaluated to determine a suitable shelf life for the product.

Applications

HPLC analysis is very important during the development of the drug formulation. It is used to assess the quality of the product throughout the development process and is used in stability studies to monitor degradation impurities.

HPLC may also be used for tests which are associated with the formulated product such as content uniformity, a test to ensure that an individual dosage unit contains the correct amount of drug, and analysis of dissolution samples, a test performed on solid dosage forms to ensure that the drug is released into the body at a suitable rate.

Another use of HPLC is for the analysis of cleaning samples. Equipment used in development of both drug substance and drug product need to be cleaned satisfactorily before they can be used for another drug development project and testing is carried out to ensure that the cleaning is sufficient to remove traces of the drug. The samples generated may be tested by HPLC.

Launch phase

If the clinical trials show good results and all the development is complete the drug is registered with regulatory bodies such as the Food and Drug Administration (FDA) in the US and European Medicines Agency (EMEA) in Europe. All the data associated with the drug is compiled in an application document, New Drug Application (NDA) for the US and Marketing Authorisation Application (MAA) for Europe. If the drug is approved then the drug goes into full scale manufacturing and is available for patients.

Manufacturing

The manufacturing facility will routinely generate large batches of drug batches of both the drug substance and the formulated product. Also, stability studies are carried out on the material produced from the manufacturing plant.

Applications

Once the drug is approved HPLC is used routinely by QA/QC laboratories for monitoring the quality of the manufactured drug substance and formulated product. Assay and impurity determination is common but it may also be used for content uniformity and dissolution analysis. Other uses of HPLC in manufacturing include analysis of environmental samples which requires the detection of the drug in effluent waste samples at trace levels.

Contents of a HPLC Analytical Method

A HPLC analytical method provides the necessary information to enable a HPLC analysis. The purpose of the analysis is to gain information on a drug or a related compound. The information may be qualitative, in which case the retention time is always constant under identical chromatographic conditions, hence a peak can be identified by comparing retention times, or quantitative, in which case the area and the height of the peak is proportional to the amount of the compound injected. Calibration graphs can be derived from solutions of precisely known concentration and used to determine the concentration of an unknown.

In the early stages of the drug discovery and development HPLC analytical methods may not be developed for a specific separation. Instead generic methods are used where a single set of chromatographic conditions is used for all samples. This approach is not optimised for a particular compound and peaks may co-elute but it is sufficient for this stage of development and enables a high throughput of analyses. This also enables the use of 'walk-up' instruments where chemists can analyse samples using predefined methods without lengthy HPLC method development. These instruments often have MS detection so that identification of the components in the samples is possible.

In the later stages of drug development, accurate quantitative information, particularly regarding impurities is required. To ensure that all impurities are detected methods that are specific to the separation are required. In the case of stability studies, methods need to be able to detect degradation impurities which may occur at some time in the future. This type of compound specific method development can be time consuming. In post-approval manufacturing the requirements for a quality control method are that it is robust and quick so that batches can be released as quickly as possible.

The information in the method is defined during the process of method development. At this stage the requirements are considered and a suitable stationary phase and mobile phase are selected. The chromatographic conditions are optimised to achieve the desired separation. In the case of a drug and its impurities where the structure is very similar it may be very difficult to achieve a separation. When suitable conditions have been found the HPLC analytical method comes into being. The method will then be validated. **Method validation** involves testing a number of parameters to ensure that the method can produce the same result consistently. Through use of the method over time experience of how it performs is gained and extra information may be added if necessary, e.g. system suitability testing may be fine tuned (see Chapter 7 for more information on system suitability).

All HPLC analytical methods contain the same basic information. The column, mobile phase and system parameters will be specified as a minimum. The information that may be included is shown in Table 19.

Table 19   Typical sections of a HPLC analytical method

Section	Description
Chromatographic parameters	Details the requirements for the chromatographic system.
Preparation of test solutions	Directions on how to prepare the system suitability test solutions (if appropriate), the sample solution and any calibration standard solutions.
Directions for analysis	Information on the injections to perform and the system suitability requirements.
Example chromatograms	Chromatograms provided for comparison with those obtained during analysis as a qualitative check.
Calculations	Details how to perform the calculations to obtain the desired results.

The order in which these typical sections are listed may vary. For example, HPLC analysis tests in USP monographs are listed as follows: First, the directions for the preparation of the mobile phase, then the standard preparation (and system suitability test solution if appropriate), followed by the sample preparation. Details on the chromatographic system including the system suitability requirements are then given and finally the procedure which includes the calculation.

The level of detail contained in the HPLC analytical method will depend on both the stage of drug discovery or development and also the preference of the laboratory where the analysis is being performed. Generally, in early stage drug discovery and development the HPLC method will not be very detailed, but by the time the drug is fully developed and being manufactured to be sold the HPLC method used for quality assurance is likely to be very detailed. Requirements for the level of detail in HPLC analytical methods will also vary in different laboratories. In some laboratories templates may exist so that the same format is always used when writing a method. An example of a HPLC analytical method is provided in Figure 23 (page 78).

Chromatographic parameters
The chromatographic parameters relate to the set up of the HPLC system for the analysis. This typically includes:

1. Column

2. Mobile phase

3. Mobile phase composition/gradient time table

4. Flow rate

5. Injection volume

6. Temperature

7. Detector settings e.g. wavelength

Preparation of test solutions

The typical test solutions that will be required are the system suitability test solutions, the calibration standard solution and the sample solution.

The system suitability test solutions will depend on the individual method (refer to Chapter 7). A solution containing a number of components is required for a resolution test, this may be prepared by weighing the components into a flask or sometimes the drug substance is degraded so that the resolution of the resulting impurities can be measured. Also, a dilution of the calibration standard may be required for a limit of quantification test.

The preparation of the calibration standard solution is likely to be straightforward unless there are solubility issues with the compound. The directions may just state a desired concentration (e.g. 'prepare a 0.2 mg/mL solution of...') in which case an appropriate volumetric flask is selected by the analyst. The method may be explicit in the weighing and glassware which should be used (e.g. 'weigh accurately approximately 20 mg of reference standard into a 100 mL volumetric flask'). The accuracy of the standard weighing may be checked by preparing another standard solution and comparing the response obtained for each preparation. The response factor of detector response (measured as area or height for the peak of interest) divided by the weighing should be the same for both preparations, a significant difference implies a weighing error.

The sample types that may be encountered in pharmaceutical analysis are varied. They include:

1. Solids e.g. isolated drug substances, intermediates and raw materials. Also solid dosage forms such as tablets and capsules.

2. Liquid formulated products, e.g. injectables and oral suspensions.

3. Formulations of creams, lotions and ointments.

4. Biological samples, e.g. plasma and urine.

The preparation of solids and liquids are usually straightforward but formulated products such as creams and ointments and biological samples will usually require an extraction technique. The most common method used is solid phase extraction. (SPE). This is an extraction method that uses a solid phase and a liquid phase to isolate an analyte from a solution. The same types of stationary phase are used as for HPLC. The solution is loaded onto the SPE phase, any undesired components are washed away, and then the desired analyte is washed off with another solvent into a collection tube. Alternatively, the undesired components may be retained by the SPE phase and the desired analyte washed off first. These types of sample preparations will usually necessitate the use of an internal standard calibration method (refer to Chapter 8).

Filtration of samples which contain particulates, such as solid dosage forms, is recommended to prevent problems with the HPLC system. Particulates can clog the injector needle, and will be collected on the frit of the column, contributing to the back pressure and ultimately reducing the lifetime of the column. Filtration is usually performed using a disposable syringe and filter. These filters are typically of pore size 0.2 μm and are available in different materials (e.g. PTFE, nylon) depending on whether a hydrophilic or hydrophobic system is required.

The number of sample preparations for a given analysis may be defined in the HPLC analytical method or by local laboratory procedures. It is common to prepare samples in duplicate to allow statistical comparison of the two results. In this way sample preparation errors may be detected.

It is very important that the stability of the test solutions is known so that they can be prepared and analysed in a suitable time scale. The solution stability is evaluated during the validation of the method and ideally this information should be included in the HPLC analytical method.

Directions for analysis

Typical information included in this section includes:

1. The injections to perform, e.g. calibration standard, system suitability test, blank (solvent), sample.

2. The number of injections to perform, e.g. 2 injections of each sample solution, or, 6 injections of the calibration standard are required for system suitability (refer to Chapter 7).

3. The expected retention time for components of interest, often a table of the retention times is presented including the **relative retention times** (RRT), i.e. the retention times of minor components expressed as a ratio of the retention time of the major component, the RRT of the major component being equal to 1. RRTs help to assign the identity of peaks. Small differences in retention time from analysis to analysis are expected but the RRT should remain constant.

4. The acceptance criteria for the system suitability test (refer to Chapter 7).

5. Details of the standards which should be used for calibration of each sample, e.g. the samples may be bracketed by standards injected before and after, these standards are then used for the calibration which is used to quantify this sample (refer to Chapter 8).

Injections of the calibration standard are performed regularly throughout a chromatographic run. This is to ensure that the response is consistent over the course of the analysis. The regularity of the standard injection may depend on the length of each injection. Examples could be after every four injections of sample, or perhaps every 2 hours. The regularity may be specified in the method or may be defined by local laboratory procedures. Sometimes the sequence of the injections and the number of injections to be performed are also determined by the local procedures and therefore are not included in the HPLC analytical method.

Example Chromatograms
If example chromatograms are available then the peak shape, the baseline, the solvent front, the retention times and the relative magnitude of the peaks can all be compared with those obtained during the analysis. This is a very useful way to identify any problems during an analysis.

Calculations
In this section the calculations required to convert the information in the chromatogram into analytical results are detailed. Calibration and quantification are discussed in more detail in Chapter 8.

Interpretation of the HPLC Analytical Method

The detail which is included in a HPLC analytical method will dictate how much interpretation is required by the analyst when following the method. For a very detailed method the analyst should be able to follow it easily but if the method does not include certain details the analyst will have to make some decisions about how to perform the analysis.

Local laboratory procedures
Routinely parts of the HPLC analytical method are covered by the local procedures in the analytical laboratory and thus will not be included in the HPLC analytical method. Examples are:

1. The sequence of the injections to be performed.

2. The procedure to use when preparing the mobile phase, e.g. the grades of solvents which should be used.

3. The routine use of guard columns or cartridges.

4. The implementation of wash programmes for column post analysis.

A good knowledge of the local laboratory procedures is required to ensure that all requirements are met.

HPLC column not defined or unavailable

When following a pharmacopoeia method, the column to use will be defined by the bonded phase type (e.g. octadecyl silane) and the particle size range (e.g. 1.5 to 10μm). This makes it difficult to choose which column to use since these parameters may be used to describe hundreds of columns. It is up to the analyst to select a suitable column. There may be a preference in the laboratory for a particular column to use in these situations. Some method development may be necessary to achieve the correct results following the monograph.

Another problem related to columns which may be encountered is that the column needed for a method is no longer available. In this case an equivalent column needs to be sourced. To find an equivalent column a column classification system is required. There are a number of researchers working in this area and their work has produced data for comparing columns[3,4].

Dwell volume

When using gradient methods the effects of dwell volume can lead to changes in the retention times when methods are transferred between different instruments. Ideally methods are developed to be robust to changes in dwell volume but if the change results in reduced resolution a solution will be required.

If the retention times are shorter than expected due to the difference in dwell volume then the current system has a smaller dwell volume than the previous system. A solution is to introduce a 'hold' of the initial gradient mobile phase composition at the beginning of the analysis. The length of the hold can be determined by measuring the dwell volume on the two systems and dividing the difference by the flow rate, or, it can be determined experimentally by injecting some test samples with different hold values programmed into the gradient table.

If the retention times are longer than expected due to the difference in dwell volume then the current system has a larger dwell volume than the previous system. Unless the injector has a function whereby the injection can be performed after the gradient has started (unlikely in older systems which are more likely to have a large dwell volume) the method cannot be satisfactorily run on the instrument.

The dwell volume of an HPLC system is easy to measure:

1. Remove the column from the system and use a short length of 0.010" tubing to connect the injector directly to the detector.

2. For solvent A, use HPLC grade water; for solvent B, add about 0.1% acetone to water (methanol or acetonitrile can be used instead of water).

3. Set the detector wavelength to 265nm.

4. Run a typical gradient from 0 to 100% B (e.g. 0-100% in 20 minutes at 3 mL/min flow). Record the detector signal during this gradient.

5. Print out or display the "chromatogram" from the gradient run. It should look like Figure 22. Draw the best straight line fit to the flat portion at the beginning of the plot. Draw the best straight line fit to the linear ramp of the gradient. The time at which these two lines intersect is the dwell time (t_D). The dwell volume is the product of the dwell time and the flow rate used for the test.

Figure 22 Measurement of the dwell volume

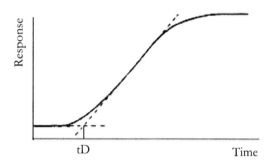

Figure 23   An example of a HPLC analytical method

Analytical Method for Determination of the Assay and Impurities of 'MiracleCure' Drug Substance by HPLC

Chromatographic parameters

Instruments	HPLC system with a UV detector			*Flow rate*	1 mL/min	
Column	Symmetry C18, 10cm, 4.6mm i.d., 3.5μ			*Injection volume*	5μL	
Mobile Phase A	25mM Potassium phosphate, pH 3			*Temperature*	30°C	
Mobile Phase B	Acetonitrile			*Wavelength*	240nm	

Gradient	Time	%A	%B
	0	80	20
	40	30	70
	40.1	80	20
	50	80	20

Preparation of test solutions

Diluent   50/50 acetonitrile/water

SST solution   Weigh approximately 2.5mg of Impurity X analytical reference standard into a 10mL volumetric flask and make to volume with diluent. Call this solution X. Weigh accurately approximately 25mg of 'MiracleCure' analytical reference standard into a 100mL volumetric flask, then pipette 1mL of solution X into the same flask. Make to volume with diluent.

Calibration standard   Weigh accurately approximately 25mg of 'MiracleCure' analytical reference standard into a 100mL volumetric flask and make to volume with diluent.

Standard check   Weigh accurately approximately 25mg of 'MiracleCure' analytical reference standard into a 100mL volumetric flask and make to volume with diluent.

LOQ solution   Pipette 1mL of the calibration standard in to a 100mL volumetric flask and make to volume with diluent to obtain a 1% solution. Pipette 1mL of the 1% solution into a 20mL volumetric flask and make to volume with diluent to obtain the LOQ solution.

Sample   Weigh accurately approximately 25mg of 'MiracleCure' drug substance sample into a 100mL volumetric flask and make to volume with diluent. Prepare in duplicate.

Procedure

Inject the SST solution, the diluent, The LOQ solution, the standard check, the calibration standard and the sample. 6 Injections of the calibration standard are required for repeatability, thereafter the calibration standard should be injected every four hours between samples and the last injection of the analysis should be of calibration standard.

The peak due to 'MiracleCure' in the SST solution should elute at a retention time in the range of 9 to 12 minutes. The peak due to Impurity X in the SST solution should elute at a relative retention time of 1.21. The resolution between the peak due to 'MiracleCure' and the peak due to Impurity X in the SST solution should be > 2.0. The tailing factor for the peak due to 'MiracleCure' in the SST solution should be <2. The %RSD for 6 injections of standard should be not more than 1.0%. There should be no interfering peaks in the diluent injection.

Use all the injections of calibration standard to construct a calibration curve for the peak due to 'MiracleCure'.

Calculate the recovery of the standard check solution, the recovery of the peak due to 'MiracleCure' in the standard check should be within the range 99.0 to 101.0%.

Calculate the recovery of the LOQ solution, the recovery of the peak due to 'MiracleCure' in the LOQ should be within the range 0.04 to 0.06%.

Calculate the assay for 'MiracleCure' in the sample.

Calculate each of the impurities in the sample disregarding impurities below 0.05% w/w.

Table of peaks with expected retention times and relative retention times:

Name	Retention time (minutes)	Relative Retention Time (RRT)
MiracleCure	10.5	1.00
Impurity X	12.7	1.21
Impurity Y	35.1	3.34

Calculations

The calculation for %RSD, resolution and tailing factor are as detailed in USP 30 <621>.

Recovery of standard check solution =

$$\frac{A_{std\ check}}{A_{cal\ std}} \times \frac{W_{cal\ std}}{W_{std\ check}} \times 100$$

Where:

$A_{std\ check}$ = Area of peak due to 'MiracleCure' in standard check solution injection

$A_{cal\ std}$ = Average area of peak due to 'MiracleCure' in calibration standard solution injections

$W_{std\ check}$ = Amount of 'MiracleCure' in standard check solution (mg)

$W_{cal\ std}$ = Amount of 'MiracleCure' in calibration standard solution (mg)

Recovery of LOQ solution =

$$\frac{A_{LOQ}}{A_{cal\ std}} \times 100$$

Where:

A_{LOQ} = Area of peak due to 'MiracleCure' in LOQ solution injection

Assay (%w/w) =

$$\frac{A_{sam}}{A_{cal\ std}} \times \frac{W_{cal\ std}}{W_{sam}} \times 100$$

Where:

A_{sam} = Area of peak due to 'MiracleCure' in the sample solution injection

W_{sam} = Amount of 'MiracleCure' in sample solution (mg)

Assay should be determined for each duplicate and the mean of the results reported.

Impurities (%w/w with respect to 'MiracleCure') =

$$\frac{A_{imp}}{A_{cal\ std}} \times \frac{W_{cal\ std}}{W_{sam}} \times Assay$$

Where:

A_{imp} = Area of peak due to the impurity in the sample solution injection

Assay   = Mean result for assay (%w/w)

Each impurity above the disregard limit should be quantified for each duplicate sample preparation and the mean for each impurity calculated and reported.

Summary

1. The three major stages in the process of creating a new drug are: drug discovery, drug development and manufacturing. HPLC is an important technique throughout all three stages.

2. A HPLC analytical method provides the necessary information to enable a HPLC analysis.

3. The contents of a typical HPLC analytical method will include: chromatographic parameters; preparation of test solutions; directions for analysis; example chromatograms and calculations.

4. Some interpretation of the HPLC analytical method by the analyst may be necessary.

References

1. W. Sneader, '**Drug Development: From Laboratory to Clinic**', Wiley, 1986.
   A comprehensive treatment of the process of creating a new drug.

2. G.L. Patrick, '**An Introduction to Medicinal Chemistry**', 3rd ed., Oxford University Press, 2005.
   A detailed textbook aimed at medicinal chemistry students but is also a very useful resource for the pharmaceutical analyst.

3. D. Visky, E. Haghedooren, P. Dehouck, Zs. Kovács, K. Kóczián, B. Noszál, J. Hoogmartens, E. Adams, , J. *Chromatogr. A*, **1101**, 103-114, 2006, '**Facilitated column selection in pharmaceutical analyses using a simple column classification system**'.
   This research group provide their column classification data on a website (www.pharm.kuleuven.be/pharmchem).

4. M.R. Euerby, P. Petersson, *J. Chromatogr. A*, **994**, 13-36, 2003, '**Chromatographic classification and comparison of commercially available reversed-phase liquid chromatographic columns using principal component analysis**'
   This research group has a collaboration with Advanced Chemistry Development (ACD/Labs) to provide a column selection tool (www.acdlabs.com).

Further reading

M.S. Lee, E.H.Kerns, Mass Spectrometry Reviews, **18**, 3-4, 187-279, 1999, '**LC-MS Applications in Drug Development**'.

W.A. Korfmacher, Drug Discovery Today, **10**, 20, 1357-1367, 2005, '**Principles and Applications of LC-MS in New Drug Discovery**'.

Notes

Performing HPLC Analysis

To perform a HPLC analysis it is necessary to set up the instrumentation and prepare the appropriate test solutions following the HPLC analytical method. In this chapter the procedure for following a method is described step-by-step using the example of a HPLC analytical method provided in Figure 23 (page 78). This guide can be modified as required to provide the analyst with a checklist to use when performing HPLC analysis.

Step 1 – Collect the required materials

It is necessary to read through the method carefully both to familiarise with the contents and also to ensure that all the items specified are available. Ideally this should be done several days before performing the analysis so that any materials that are not available can be ordered. In Table 20 the items that are typically required for a HPLC analysis are detailed alongside those required for the example.

Table 20   Required materials for HPLC analysis

Required materials	Example
Column	Symmetry C18, 5 micron, 15cm x 4.6mm i.d.
Mobile phase solvents	Water; Acetonitrile
Mobile phase reagents	Potassium phosphate; Orthophosphoric acid/ammonium hydroxide (to adjust pH if required)
Solvent for standard and sample preparation	Water; Acetonitrile
Analytical instruments	PH meter; Balance
Standard(s) (if required)	'MiracleCure' analytical reference standard; Impurity X analytical reference standard
Sample	'MiracleCure' drug substance sample to be analysed
Glassware etc. for preparation of test solutions	100mL volumetric flasks; 1mL volumetric pipette
Suitable HPLC system including the detector specified in the HPLC analytical method	HPLC system with UV detector

HPLC columns which have the same bonded phase do not necessarily have the same selectivity, as discussed in previous chapters. Therefore, the column specified in a method cannot be replaced with any column of the same bonded phase, e.g. a Hypersil C18 may not give the same chromatography as a Symmetry C18 for a given mixture. If the column specified in the analytical method is not available then extra work may be required to prove that the replacement column is suitable. Replacement of columns was discussed in Chapter 6.

For HPLC analytical methods that are performed regularly it is a good idea to dedicate an individual column to the method. This means that the column is only exposed to the mobile phase system for that method and reduces the risk of a change in column chemistry that could affect the chromatography. This is a common practice in many pharmaceutical analysis laboratories where routine analysis is performed. Several columns are usually dedicated to the method and new columns are ordered regularly to ensure that a spare is always available. When walk up HPLC systems are used with generic HPLC analytical methods an adequate supply of spare columns is essential.

Action

Collect all the necessary materials and order any that are unavailable.

Step 2 – Record the analysis

When working in a GMP/GLP environment it is essential that the analysis be recorded at the same time as it is performed. The recording of the analysis needs to be such that every action performed by the analyst can be traced. For non-GMP/GLP analysis it is advisable to follow similar principles of data recording since this will give confidence in the results obtained and also make it easier to troubleshoot if required.

The format of the 'write up' will depend on the way of working in a given analytical laboratory. The options include: handwritten notebook entries, electronic recording of data or perhaps a combination of the two. The Chromatography Data Systems (CDS) attached to most modern HPLC systems allow the electronic capture of most of the data associated with a HPLC analysis plus they are able to calculate the results and generate reports. They may be linked to other data handling systems, e.g. Laboratory Information Management Systems (LIMS).

Typical information recorded for a HPLC analysis will include: name of the analyst; date the analysis was performed; brief description of the aim of the analysis; details of weighings; identifiers for the reference standards and samples and an identifier for the HPLC system used. Prior to performing any practical work in the laboratory any available information should be detailed in the 'write up'.

Action

Start the write-up of the analysis by assigning a notebook or electronic identifier, and recording the available information. Continue to record the required information when performing the remaining steps.

Step 3 – Prepare the mobile phase

To determine the amount of mobile phase that will be required for an analysis the total number of injections needs to be calculated. The total number of injections is then multiplied by the run time for one injection and the flow rate to give the amount of mobile phase required. However, extra mobile phase is always needed to enable the set up of the HPLC system, and to allow for extra injections should the need for these arise, so an additional amount is usually added on for this.

The injection sequence

The injection sequence for the analysis needs to be determined in order to calculate the total number of injections required. This will depend on the number of samples for analysis and also the requirements for the system suitability test and standard injections. Using the example method on page 78 the system suitability test requires 1 injection of the system suitability test solution, an injection of the blank, and an injection of the LOQ test solution, an injection of the standard check and 6 injections of the standard at the beginning of the chromatographic run. If there are 5 samples to be analysed in duplicate this adds another 10 injections (5 was selected at random as the number of samples for illustrative purposes). In addition standard injections are required every 4 hours between the samples and one standard injection at the end. All this information is converted into an injection sequence in Table 21.

Table 21 Injection sequence to analyse 5 samples using example HPLC method

Injection number	Test solution to be injected
1	SST solution
2	Blank
3	LOQ solution
4	Standard check solution
5	Calibration standard
6	Calibration standard
7	Calibration standard

Injection number	Test solution to be injected
8	Calibration standard
9	Calibration standard
10	Calibration standard
11	Sample 1/1
12	Sample 1/2
13	Sample 2/1
14	Sample 2/2
15	Calibration standard
16	Sample 3/1
17	Sample 3/2
18	Sample 4/1
19	Sample 4/2
20	Calibration standard
21	Sample 5/1
22	Sample 5/2
23	Calibration standard

Mobile phase required

In this example there are 23 injections in the sequence and the flow rate is set at 1mL/min. Therefore the minimum amount of mobile phase required is $23 \times 50 \times 1 = 1150$mL. Preparing 1.5L of mobile phase will give a sufficient quantity for system set up and a contingency for extra injections.

For an isocratic method, where the mobile phase is premixed, the correct proportions would be prepared in a single mobile phase reservoir and placed on the HPLC system. However, the example is a gradient method and the amounts required for mobile phase A and B need to be considered. Typically gradient methods include a re-equilibration delay after an injection and before the next one, this involves a flow of the mobile phase at the starting conditions of the gradient (in the example the mobile phase composition of 80% A and 20% B is held for 10 minutes after each injection). As a result, more of mobile phase A is required than mobile phase B.

Using the example and assuming that the mobile phase is premixed, 20% of the run time for an injection is at the starting conditions and therefore using mobile phase A. The other 80% of the run time is split evenly between mobile phase A and B. Therefore, A and B are required in the approximate proportions of 60% A and 40% B, or expressed as volume, 900mL A and 600mL B to give 1.5L in total. If on line mixing is used then the proportions of A and B relate to aqueous phase and organic solvent respectively and approximate amounts of each need to be calculated in a similar way as for the previous premixed example. If in doubt, it is best to make more mobile phase than will be needed, it is better to have mobile phase left over at the end of the analysis than to run out during the chromatographic run. Adding new mobile phase during an analysis is not recommended. The small differences between preparations may be enough to alter the retention times during the analysis and also cause irreproducibility of the baseline from injection to injection. For analyses where on line mixing is being used and pure solvent is being mixed by the HPLC pump to form the correct mobile phase, the reservoirs may be topped up using more solvent from the same batch without adverse effects providing sources of contamination are avoided.

Sometimes the total number of samples for analysis is not known. For HPLC systems where individual injections of samples are performed as and when they are available, such as walk up analysis systems, then a stock amount of mobile phase is usually prepared at regular intervals and replenished as required.

Mobile phases should be prepared following any appropriate local procedures regarding grades of reagents to use, filtering, degassing and premixing. Mobile phase preparation is discussed fully in Chapter 3.

Action

Calculate the required quantity of mobile phase.
Prepare mobile phase and mix, filter, degas as required.

Step 4 – Set up HPLC system

The first step when setting up a HPLC system is usually to flush the system with the mobile phase which will be used for the analysis. Before this can be done it is important to know what the system has been used for previously. The liquid which is present in the system needs to be miscible with the new mobile phase. If a normal phase separation has been performed previously and the new mobile phase is reversed phase then the system will need to be flushed with a number of different solvents of increasing polarity until the new mobile phase can safely be introduced to the system. In practice, normal phase HPLC is not used very often in pharmaceutical analysis. If normal phase is used regularly, e.g. chiral separations, a HPLC system is often dedicated to normal phase analysis since changeover from normal phase to reversed phase and vice versa is time consuming and it usually takes a long time for the system to equilibrate to the new conditions afterwards.

Flush the pump

To flush the HPLC system with the new mobile phase, first place the new reservoirs onto the system and insert the tubing. Then switch on the pump (and degasser if installed). Most pumps have a purge valve which allows liquid going through the pump to go to waste. The purge valve should be set to waste and a high flow rate set to flush the new mobile phase through the pump. This process is referred to as priming the pump.

Flush the injector

The next stage is to flush through the injector and tubing up to the column inlet. Close the purge valve to direct the mobile phase through the HPLC system and turn on the flow using a vessel to collect the mobile phase flowing through at the column inlet (since the column is not yet installed). The injector may have a purge function, this will depend on the design. If it does then the injector may be purged using the new mobile phase. Now turn off the flow, the column can be installed.

Connect and flush the column

The column is connected using the fittings as described in Chapter 4. These will be either stainless steel or polymeric and may be finger tight or require the use of a spanner. In either case be careful not to over tighten. Ensure that the direction of flow is the same as the arrow indicated on the column. The flow of mobile phase is always passed through the column in the same direction so that impurities which are trapped on the inlet frit of the column are not released into the detector in a subsequent analysis. If a column is being used for the first time and there is no arrow marked on it, then it is a good idea to draw an arrow on the label so that it will always be used in the same direction in future analyses. The flow is now switched on at the flow rate specified in the method and a small aliquot collected at the end of the column. (Most modern systems will increase the flow rate slowly until it reaches the programmed value so as to avoid creating sudden pressure at the head of the column.) This flushes the column and prevents any air in the column being passed on to the detector. The flow is then switched off. The column outlet is connected so that the flow of mobile phase will travel through the detector and then to waste.

Turn on and equilibrate the entire system

The flow rate is switched on at the value defined in the method (1mL/min in the example). A constant pressure should be observed. The other parts of the HPLC system are switched on and the values set to those given in the method. For the example, the wavelength of the detector is set to 240nm and the column oven temperature to 30 degrees. These values may be set up on the instruments or may be controlled through a CDS. The system needs to equilibrate before it can be used for analysis. The length of time that is required for equilibration will vary for different systems and methods. Some bioanalysis methods require very long column conditioning prior to analysis. The output from the detector is usually monitored and when the baseline is level and steady the system should be ready.

If the HPLC system has needle-wash and seal-wash reservoirs these should be filled with an appropriate solvent, also the waste vessel should be emptied.

Action

Flush the HPLC system with the new mobile phase.
Attach the HPLC column.
Adjust the detector and column oven to the required setting.
Fill reservoirs of needle-wash etc and empty the waste receptacle.
Set the flow rate to the desired value.
Monitor the output from the detector.

Step 5 – Prepare the test solutions

The test solutions are prepared as directed in the HPLC analytical method. In the example a SST (system suitability test) solution, a calibration standard solution, a standard check solution, a LOQ solution and samples (in duplicate) are prepared using a diluent which is made up of 50% acetonitrile and 50% water.

Sample preparation

Since the sample in the example is an analysis of a drug substance the sample preparation is straightforward. Potential samples which may be encountered in pharmaceutical analysis were listed in Chapter 5, these may include extraction procedures and filtration of the sample. Since these preparations are developed for a specific sample the directions should be included in the HPLC analytical method. If solid phase extraction (SPE) is required then the correct phase and the eluting solvents should be specified. If samples are filtered then a suitable filter membrane should be specified. The most common choices of membrane for general purpose filtration are nylon or polypropylene for aqueous samples and PTFE for hydrophobic samples.

Filling the HPLC vials

The test solutions for HPLC analysis are injected from vials as discussed in Chapter 4. Appropriate vials and lids for the vials should be selected for the analysis. The vials may be filled using Pasteur pipettes, using a new pipette for each solution to avoid cross contamination. Labelling the vials is recommended to keep track of the contents of each. Although it is possible to inject from any vial in any order using modern programmable autoinjectors, in practice it is easier to arrange the vials in order of injection. The injection sequence, which was determined to calculate the mobile phase requirements, should be used to arrange the vials into the correct order. Depending on the injection volume, multiple injections may be taken from one vial (e.g.

calibration standard injections) or if preferred a single injection may be taken from multiple vials.

Action

Prepare the test solutions as directed in the method.
Transfer aliquots of each test solution into appropriate HPLC vials.

Step 6 – Start the chromatographic run

Priming injections

For most HPLC analyses a priming injection is required prior to the analysis. This injection will usually result in a slightly different chromatogram to subsequent injections of the same solution. The reason for this is not fully understood but is probably due to the presence of (at least) two types of active sites on the column for interaction with the analyte molecules. One of these sites equilibrates much more slowly than the other and is gradually saturated with the analyte in the first few injections thus stabilising the response for further injections. For this reason it is common practice to perform one or more 'test injections' before starting the analysis injection sequence. The SST solution or the calibration standard is commonly used for the test injection. The test injections may be programmed into the injection sequence if the run will be unattended, e.g. it is started at the end of a working day.

Program the analysis

The autoinjector needs to be programmed with the necessary information to perform the injection sequence, i.e. the vials to inject, the number of injections and the injection volume. This may be done on the HPLC instrumentation using a direct interface with the autoinjector, or may be programmed into the CDS. CDS software may be used to process the data and report the results automatically, the appropriate options are selected and the chromatographic run is started.

Action

Set up the injection sequence on the autoinjector or through the CDS as appropriate and start the run.

Step 7 – Generate the analytical results

On completion of the chromatographic run a series of chromatograms are obtained. These should be visually inspected to make sure that the run has been completed satisfactorily and that there are no obvious problems. If the HPLC analytical method contains example chromatograms then these provide a good qualitative comparison at this stage.

Integration

The next step is to assess the results of the system suitability test for a more definitive measure of a satisfactory chromatographic run. Integration of the peaks in the chromatograms is required for some of the system suitability testing therefore this process will be explained. Integration is a mathematical process whereby the area under the peak is measured, it provides a quantitative measure of the size of the peak. To measure the area it is necessary to draw a baseline under the peak which defines which area will be included in the measurement. Integration is easily performed using computers although difference software packages have slightly different approaches as to how it is performed. Integration events may be used to adjust the baseline. See the reference section at the end of the chapter for a textbook about integration in chromatography[1].

Some basic guidelines for integration are listed below:

1. The integration of the peak should result in a continuation of the baseline under the peak.

2. The integration should be consistent for all the injections throughout a chromatographic analysis, and the peaks in the samples and in the standards should be integrated in the same way.

3. Overlapping peaks where the minor peak elutes on the tail of the major peak (a common situation in impurity analysis) should be integrated using a tangential skim as shown in Figure 24 (a), not by dropping a perpendicular to the baseline as shown in Figure 24 (b).

Figure 24 Integration of a small peak on the tail of the major peak

(a)   (b)

4. For larger peaks it may be more appropriate to drop a perpendicular to the baseline as shown in Figure 25 (a), rather than using a tangential skim as shown in Figure 25 (b). The decision is based on the judgement of the analyst.

Figure 25 Integration of overlapping peaks

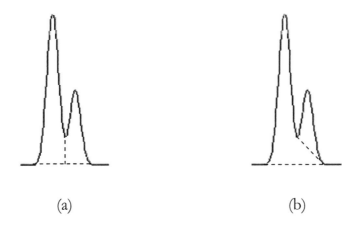

(a)   (b)

5. The threshold value will determine the peak start and peak end. If the threshold is set correctly then the baseline will continue under the peak as shown in Figure 26 (a). Figure 26 (b) shows an incorrect threshold setting.

Figure 26 Integration threshold

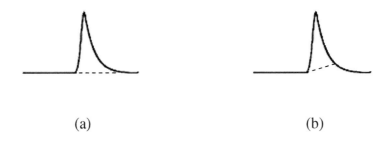

(a)   (b)

System Suitability Testing

When performing integration the injections required for the system suitability testing may be given priority. These results should be calculated following the instructions given in the method. Comprehensive system suitability requirements are detailed in the example method. Chapter 7 contains more details on the purpose of system suitability testing and the parameters that are measured. A pharmacopoeia is normally referenced for the calculations. If the system suitability does not pass then the results for the analysis are not valid and the results for the samples should not be quantified. When the reason for the failure is established and corrected the analysis will need to be repeated.

Calibration and quantification

When the system suitability has passed and the integration is satisfactory for all injections the results for the analysis may be quantified. The calculations are performed as detailed in the method (refer to the example method on page 78). The write up of the analysis may then be completed. In a late stage development or manufacturing environment the write up of the analysis will need to be checked and approved to comply with GMP requirements.

> **Action**
>
> Inspect the chromatograms obtained for the analysis.
> Integrate the chromatograms.
> Calculate the system suitability results.
> Calculate the results for each sample.

Step 8 – Housekeeping

After the analysis the HPLC system and column should be flushed with a solvent mixture which does not contain any buffer salts. This is very important when using inorganic buffer solutions, since these can precipitate out of solution and damage the components of the HPLC system, e.g. pump heads, and the column. For reversed phase methods the wash solvent is usually a mixture of water and the organic solvent used in the analysis. In the example water and acetonitrile would be used. A mixture of 50% of each is suitable, the proportions may be adjusted as desired. This wash may be programmed into the analysis so that it occurs automatically when all the injections are complete.

After the flush is complete the system is available for use and in a busy laboratory it may be required immediately. In this case the mobile phase, the test solution vials and the column should be removed and the waste vessel should be emptied. The test solutions are usually retained until the results of the analysis are processed in case re-analysis is required. They should be adequately labelled and stored safely until they can be disposed of. The column is returned to the store when the analysis is complete and if a column log system is in use this should be filled in.

> **Action**
>
> Flush the HPLC system and the column after use.
> Empty all waste vessels.
> Dispose of the mobile phase, the test solutions and the autoinjector vials when analysis is complete.
> Return column to store.

Using the checklist

94

This suggested sequence of events will enable an analyst to follow a HPLC method. The checklist is not exhaustive, it is only intended as a guide. With experience the analyst may modify the order of events to suit their own way of working. An example of this is where a method involves a time consuming sample preparation step: it may be a better use of time to begin with sample preparation and then attend to setting up the HPLC system and standard preparation. Thus the efficiency of the analysis is maximised.

Summary

A checklist to perform a HPLC analysis is as follows:

Step 1 – Collect the materials required
Collect all the necessary materials and order any that are unavailable.

Step 2 – Record the analysis
Start the write-up of the analysis by assigning a notebook or electronic identifier, and recording the available information. Continue to record the required information when performing the following steps.

Step 3 – Prepare the mobile phase
Calculate the required quantity of mobile phase.
Prepare mobile phase and mix, filter, degas as required.

Step 4 – Set up the HPLC system
Flush the HPLC system with the new mobile phase.
Attach the HPLC column.
Adjust the detector and column oven to the required setting.
Fill reservoirs of needle wash etc and empty the waste receptacle.
Set the flow rate to the desired value.
Monitor the output from the detector.

Step 5 – Prepare the test solutions
Prepare the test solutions as directed in the method.
Transfer aliquots of each test solution into appropriate HPLC vials.

Step 6 – Start the chromatographic run
Set up the injection sequence on the autoinjector or through the CDS as appropriate and start the run.

Step 7 – Generate the analytical results
Inspect the chromatograms obtained for the analysis.
Integrate the chromatograms.
Calculate the system suitability results.
Calculate the results for each sample.

Step 8 – Housekeeping
Flush the HPLC system and the column after use.
Empty all waste vessels.
Dispose of the mobile phase, the test solutions and the autoinjector vials when analysis is complete.
Return column to store.

The checklist can be modified as required.

References

1. N. Dyson, 'Chromatographic Integration Methods', 2nd ed., Royal Society of Chemistry, 1998.

Notes

Chapter

7

System Suitability

An important part of any HPLC analytical method is the system suitability test. The purpose of this is to ensure that the HPLC system is performing as expected. The test measures the performance of the HPLC system as a whole, including the mobile phase, pump, injector, column, detector, data processor and the operator. A satisfactory system suitability test provides assurance in the quality of the results produced from the analysis. This is particularly important for quantitative analysis.

System Suitability Criteria

The system suitability test (SST) is made up of a number of measurable criteria which when combined provide assurance that the system is performing satisfactorily. The typical criteria which may be applied are detailed in Table 22. The criteria which will be applied for a particular method will depend on what is appropriate for that method, therefore not all the criteria in Table 22 will necessarily be applied.

Table 22   System suitability criteria

Criteria	Description
Retention	A measure of the time at which the peak of interest elutes
Injection repeatability	A measure of the reproducibility of the system throughout the chromatographic analysis
Resolution	A measure of how well the peaks in a chromatogram are separated
Tailing	A measure of the asymmetry of a peak
Efficiency	A measure of the dispersion of a peak
Capacity factor	A measure of the how well the analyte is retained on the column

The system suitability measurements are performed at the time of the analysis and the necessary injections are included in the injection sequence. Every time a chromatographic run is carried out system suitability must be evaluated. The system suitability testing for a given method will be designed during the method development and method validation phases. System suitability testing is included as a validation parameter in the ICH guidelines on analytical method validation[1]. A cleverly designed system suitability test should get the most information out of a minimum number of injections. The requirements will be set in such a way that they will be met easily if the method is working properly but will fail if there is a method problem.

A full discussion of system suitability and the requirements associated with it are contained in pharmacopoeias. The most important pharmacopoeias worldwide and therefore those most likely to be followed are the United States Pharmacopeia[2] (USP), The European Pharmacopoeia[3] (EP) and the Japanese Pharmacopoeia[4] (JP). When performing an analysis from a monograph contained in any of these the requirements for system suitability stated must be adhered to. Since pharmacopoeias are approved by regulatory authorities the system suitability criteria stated in these are used as a guide for all HPLC methods used in pharmaceutical analysis. Although efforts to harmonise the USP, EP and JP are ongoing, there are slight differences between these in the calculation of the system suitability criteria. These are highlighted in the following discussion of each criterion.

Retention

The time at which the peak of interest elutes from the HPLC column should be constant for a given method. Therefore a deviation from the expected retention time may indicate a problem with the system. In practice a range may be specified within which it is known that the method will perform satisfactorily. This information will be gathered during method development and method validation. For gradient methods there is the added complication of different dwell volumes on different instruments and alterations may have to be made to adjust for this. This was discussed in detail in Chapter 6. The comparison of the retention time of the peak of interest for injections throughout the course of the chromatographic run may also be included in the system suitability test to ensure that the performance has remained consistent throughout the analysis.

Alterations to the chromatographic conditions to adjust the retention time are allowed by the pharmacopoeias and the flow rate may be defined in terms of the retention time in some monographs.

Injection Repeatability

Measuring the repeatability of the injections carried out during a chromatographic analysis ensures that the performance of the system is consistent. Repeatability is usually measured by calculating the relative standard deviation (%RSD). The standard deviation for a set of data points provides a measure of the spread of the individual data points. Converting this value into a percentage of the mean for the dataset enables the comparison of different sets of data. What this means for the nonstatistician is that whereas the standard deviation is only meaningful when inspecting the data it relates to, the %RSD gives an indication of the spread of the data points for any set of data. Injection repeatability is a measure of the injection **precision**.

$$\%RSD \quad = \quad \frac{\text{Standard deviation}}{\text{Mean}} \quad x \quad 100$$

To assess injection repeatability the amount being injected should be the same for each injection. Therefore the peak area is measured for a set of injections of the same solution. Usually an external standard is used for this purpose. The value obtained for %RSD depends on how well the injector on the HPLC system can reproducibly inject the solution, modern injectors are capable of very good injection repeatability. Values which would be considered to be of acceptable precision are < 1%.

The number of injections to be performed for injection repeatability is usually six. The USP requires five injections if the requirement for %RSD is 2.0% or less and six injections if the requirement for %RSD is > 2.0%. In the EP and the JP the number of injections is usually specified in the monograph. The injections may be performed at the beginning of the chromatographic analysis or throughout the run. The advantage of injecting all six prior to any sample injections is that a pass result gives confidence in the system performance. An example of a typical set of results for injection repeatability are given in Table 23

Table 23   An example of repeatability determination

Injection number	Area response for analyte peak in standard
1	451677
2	450767
3	447653
4	452556
5	449336
6	448762
Mean	450125.2
Standard deviation	1861.3
%RSD	0.41

Relative standard deviation is used as the measure of repeatability in all three pharmacopoeias (USP, EP and JP). The acceptance criteria for injection repeatability in the pharmacopoeias are given in the individual monographs. In the EP there is a rather complicated system for calculating the RSD_{max} based on the upper limit given in the definition of the monograph, a student's t-test and the number of injections performed, which is only applicable to assay determinations. For assay determinations a typical criterion may be 'no more than 1.0%', and for impurity determinations a typical criterion may be 'no more than 3.0%'. The reason that impurity determinations often have higher values of %RSD for acceptance criteria is that impurities are usually present at very low concentrations and it may be difficult to reproducibly inject at this level. The value will depend on the individual method and the difficulties associated with it.

Usually a local laboratory policy will exist documenting the acceptance criteria for injection repeatability and also the number of injections which should be performed for the different types of methods.

Resolution

Resolution measures how well two peaks are separated and is one of the most important system suitability parameters since the purpose of a chromatographic method is to separate peaks. An example chromatogram is shown in Figure 27. Peak 1 and peak 2 are not resolved but peak 3 and peak 4 are fully resolved.

Figure 27   Example chromatogram showing resolved and unresolved peaks

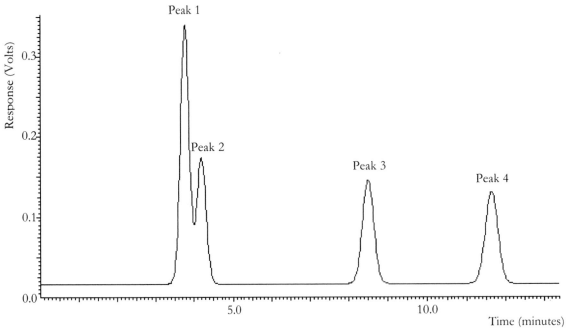

The resolution of these peaks is measured to give a numerical value of how much they are resolved. The equation used to measure resolution, as defined by the USP, is as follows:

$$R = \frac{2(t_2 - t_1)}{W_2 + W_1}$$

where t_1 and t_2 are the retention times of the two peaks on which resolution is being measured and W_1 and W_2 are the width of the peaks at the baseline measured as shown in Figure 28 for peak 3 and peak 4.

Figure 28 Measurement of resolution

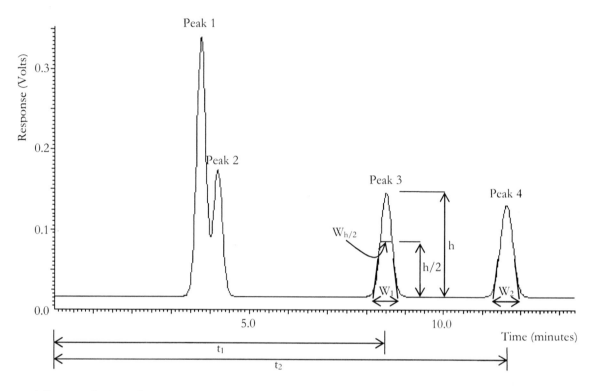

The resolution for peak 1 and peak 2 in Figure 28 is equal to 0.8 and the resolution for peak 3 and peak 4 is equal to 2.5. A value of greater than 1.5 is considered to be baseline separated. Typically resolution is measured on the **critical pair**. This is the pair of peaks which elute closest together in the separation. If these are resolved adequately then all the other peaks in the separation will also be resolved. The critical pair is identified during method development and method validation. The compounds used in the system suitability test need to be available for the preparation of the test solution throughout the life of the drug. Therefore the peaks which will be used for the resolution test need to be selected with care.

Measuring the peak width at the base of the peak can be problematic. In Figure 28 peak 3 and peak 4 are used to illustrate the measurements used for determination of resolution. If peak 1 and peak 2 were used it would be difficult to draw the tangent necessary to measure the peak width. Often peak width at half height is used for resolution measurements as shown for peak 3. The equation for resolution using peak width at half height is as follows:

$$R = \frac{1.18(t_2 - t_1)}{W_{2h/2} + W_{1h/2}}$$

This calculation for resolution is common to the USP, EP and JP although the terminology and notations used are not identical.

Tailing Factor

The tailing factor, also known as the asymmetry factor, measures the symmetry of a peak. As columns age the peak shape usually deteriorates and may lead to problems with integration. Including a measure of the peak tailing in the system suitability test enables monitoring of the column as it ages. There are other causes of peak tailing, e.g. the peak tailing caused by the interaction of residual silanol groups on reversed phase bonded phases with basic compounds has been mentioned previously. In the context of system suitability the tailing is monitored to ensure that the system is always suitable for the method. The tailing factor is determined using the following calculation:

$$T = \frac{W_{0.05}}{2f}$$

where $W_{0.05}$ and f are defined in Figure 29.

Figure 29 Measurement of tailing factor

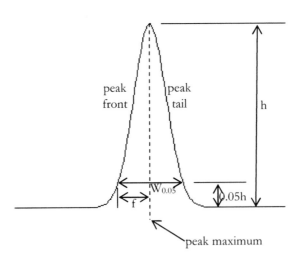

A typical acceptance criterion would be that it should not be greater than 2. This calculation for tailing is common to the USP, EP and JP although the terminology and notations used are not identical.

Efficiency

Column efficiency, also known as plate count, is a measure of the dispersion of a peak. Narrow peaks take up less space in the chromatogram and thus allow more peaks to be separated. They are also easier to integrate since they give better resolution and less overlapping. Efficiency is usually explained using the concept of theoretical plates. This model supposes that the column contains a large number of separate layers. Separate equilibrations of the sample between the stationary and mobile phase occur in these plates. The analyte moves down the column by transfer of equilibrated mobile phase from one plate to the next. It is important to remember that the plates do not actually exist, they are a means to help understand the process at work in the column. They also give a measure of column efficiency by stating the number of theoretical plates in a column, N. A high value for efficiency indicates that more peaks can be separated. The number of plates will increase with the length of the column. The calculation for efficiency is related to the peak width and is as follows:

$$N = 16 \left(\frac{t}{W}\right)^2$$

where t is the retention time of the peak of interest and W is the peak width at the base (as shown in Figure 28). As for resolution the measurement may also be performed using the peak width at half height:

$$N = 5.54 \left(\frac{t}{W_{h/2}}\right)^2$$

where $W_{h/2}$ is the peak width as half height. It can be seen that N is related to the analyte peak and thus the columns behaves as if there are different numbers of plates for each solute in a mixture. When the peak width increases resulting in broad peaks, it is due to band broadening. There are a number of different reasons for band broadening:

1. The path taken by the analyte molecules through the column varies due to chance. Some molecules will travel in a fairly straight line whereas others will undergo several diversions. The effect of this is that not all the molecules will elute at the end of the column at exactly the same time.

2. Sample molecules in a solvent will spread out without any external influence due to molecular diffusion.

3. Analyte molecules travel from the moving mobile phase to the surface of the particle, through stagnant mobile phase in the pores to the internal surface on the packing. It interacts with the stationary phase and then is transported back to the moving mobile phase. This process is referred to as **mass transfer** and not all molecules will experience mass transfer in an identical way therefore band broadening will occur.

4. The mobile phase travels in a **laminar flow** between the stationary phase particles, the flow being faster in the centre than near a particle. Thus some molecules travel more quickly than others. This flow distribution is reduced by ensuring that the particles in the packing have a narrow particle size distribution.

5. The tubing in the HPLC instrumentation contributes to band broadening, this is known as the extra-column effect.

Efficiency has been referred to previously in Chapter 2 when discussing particle size in HPLC columns packings. Smaller particle sizes give better efficiency because the diffusion paths are shorter allowing solutes to transfer in and out of the particle more quickly and thus reducing band broadening.

A typical acceptance criterion for efficiency would be > 2000. Although the value for new columns would usually be very much higher than this (values in the tens of thousands) the system suitability acceptance should be based on a value which indicates that the efficiency is no longer sufficient for the separation. These calculations only apply to isocratic separations. For gradient methods the peak width remains fairly constant throughout the run due to the changing mobile phase composition and therefore the value for N would appear to increase with retention time. A more useful measure of the column efficiency would be the peak width at half height for the analyte. Monitoring this value could provide a measure of when the column efficiency is no longer sufficient for the separation. Resolution depends indirectly on efficiency and therefore if resolution is a parameter in the system suitability test then a measure of efficiency is already included.

The calculation for efficiency using the peak width at half height is common to the USP, EP and JP although the terminology and notations used are not identical. However, due to a slight difference in rounding, the constant in the equation is 5.54 in the USP and EP but is 5.55 in the JP.

Capacity Factor

The capacity factor (k'), also known as the retention factor (k) and the mass distribution ratio (D_m) is a measure of where a peak of interest is located with respect to the void volume (i.e. the elution time of non-retained components). It is the ratio of the time that an analyte spends in the stationary phase to the time it spends in the mobile phase. The higher the value the more retained the compound. The calculation for capacity factor is as follows:

$$k' = \frac{t - t_0}{t_0}$$

where t is the retention time of the peak of interest and t_0 is the time taken for non-retained components to elute. Typically the requirement is that the capacity factor is greater than 2. This ensures that early eluting peaks are suitably resolved from the solvent front. Since the capacity factor depends on the retention time it provides no added value as a system suitability test if there are already requirements relating to retention time. It may be useful if it is known that an early eluting peak is less well retained as the column ages and therefore the capacity provides a measure of when the column is no longer suitable for retention of that analyte. It is not usually required in USP and JP monographs but may be required by some EP monographs.

Method Adjustments to Achieve System Suitability

If a system suitability test fails then action has to be taken to discover what the problem is and how it can be resolved. The problem may be due to the preparation of the mobile phase or test solutions or may be related to the HPLC instrumentation or the column. When a method is used routinely experience of how that method performs is acquired and troubleshooting a system suitability failure may be straightforward. However it can be difficult to ascertain the nature of the problem for unfamiliar methods. Each of the system suitability parameters are discussed in Table 24 with potential sources of failure.

Table 24 System suitability failures

Criteria	Comments
Retention	Small differences in retention time may occur when changing a method from one system to another which do not impact on the quality of the results. In isocratic methods the flow rate may be adjusted to achieve the required retention time if all other system suitability parameters pass. If other system suitability parameters are failing then there may be a bigger problem which needs attention. For gradient methods, differences may be due to dwell volume differences and these should be investigated (refer to Chapter 6).
Injection repeatability	The most common causes of injection repeatability failure are related to maintenance of the injection assembly. A worn sampling syringe or bubbles in the solvent used to flush the syringe can lead to poor reproducibility of injections. Unsuitable caps for the test solution vials may also be a source of problems and, if the last injection is affected where multiple injections are taken from one vial, there may have been insufficient material in the vial.
Resolution	Replacing the column is probably the best course of action if the resolution test fails providing that the mobile phase composition is not suspect and the dwell volume of a gradient method has been considered.
Tailing	As for resolution, increasing tailing is usually a sign that the column is degrading with time.
Efficiency	As for resolution.
Capacity factor	As for resolution.

Alterations to the chromatographic conditions to achieve system suitability are allowed by the pharmacopoeias. The EP has a very detailed section on what changes are allowed to achieve system suitability, including absolute limits for each type of operating condition, whereas the JP and USP list the operating conditions which may be altered without giving absolute values.

System Suitability in Practice

The system suitability testing may include other criteria which are appropriate for a particular method in addition to those discussed previously. Examples are:

1. Assessment of any peaks in an injection of the diluent (blank) especially at the retention time of the main component. This test may be included for impurities methods.

2. Assessment of the limit of quantification (LOQ). May be used for impurities methods to ensure that the disregard limit (the level at which a peak is deemed to be too low to allow quantification) is appropriate for the analysis.

3. Assessment of the limit of detection (LOD). Similar to above, ensures that the LOD determined for the method is suitable for analysis.

4. If a chromatogram is provided in the HPLC analytical method the system suitability requirements may include a clause to compare the chromatogram obtained for the analysis with that in the method and to ensure that they are qualitatively similar.

5. A comparison of the response factor (detector response divided by the amount in the solution) of standards which have been prepared from separate weighings may be included in the system suitability testing as a check on the accuracy of the weighing. This may be referred to as standard verification or standard recovery.

The extent of system suitability testing may also vary with the stage of development for the analyte drug. In the discovery stage an individual HPLC analytical method for a specific analyte may not exist, instead a generic method is used for all analytes, in this case the system suitability testing will also be a generic approach and may not relate directly to the analyte. In the later stages of drug development and during manufacturing quantitative results are required and it is essential to ensure that methods are performing correctly. Therefore more elements of system suitability may be included in the HPLC analytical method.

The details of the required system suitability testing for a given method should be included in the HPLC analytical method including directions about which solutions and injections to use for each criterion. Also, local laboratory procedures for chromatographic analysis may specify requirements for system suitability testing.

Modern chromatography data systems will perform system suitability measurements automatically and can usually be set to report to the required pharmacopoeia.

Bioanalytical Methods

Bioanalytical methods determine the amount of drug and/or metabolites present in samples derived from biological matrices such as blood, serum, plasma or urine, and are used to support clinical and non-clinical studies. In addition to system suitability these methods routinely include Quality Control (QC) samples. QC samples are prepared using the sample matrix and adding known amounts of analyte. They are used during the validation of the method and subsequently to check the performance of the method each time it is performed. If the results obtained for the QC samples are within acceptable limits it indicates that the method has performed acceptably and thus the results obtained for the study samples will also be acceptable. Acceptable limits for QC samples are discussed in the FDA guidelines[5] relating to validation of bioanalytical methods. The QC samples must be prepared and treated in the same way, and at the same time, as the study samples to ensure the validity of the results. More information about QC samples may be found in bioanalysis textbooks.[6,7]

Summary

1. System suitability ensures that the HPLC system comprising of the mobile phase, pump, injector, column, detector, data processor and the operator, is operating as expected.

2. The system suitability test is required each time analysis is performed.

3. Typical criteria used to assess system suitability include: retention; repeatability; resolution; tailing; efficiency and capacity factor.

References

1. International Conference on Harmonisation (ICH) of Technical Requirements for Registration of Pharmaceuticals for Human Use, Topic Q2 (R1): **Validation of Analytical Procedures: Text and Methodology**, 2005, www.ich.org.

2. United States Pharmacopeia (USP) 30, Chromatography <621>

3. European Pharmacopoeia (EP) 5, Chromatographic Separation Techniques, 2.2.46

4. Japanese Pharmacopoeia (JP) 15, Liquid Chromatography, 27

5. Guidance for Industry: **Bioanalytical Method Validation**, US Food and Drug Administration, Center for Drugs and Biologics, Department of Health and Human Services, 2001.

6. R. Venn, '**Principles and Practice of Bioanalysis**', Taylor & Francis Ltd., 2000.

7. A. Manz, N. Pamme, D. Iossifidis, '**A Handbook of Bioanalysis and Drug Metabolism**', Taylor & Francis Ltd., 2003.

Notes

Calibration and Quantification

The reason for the widespread use of HPLC in pharmaceutical analysis is that it enables the quantification of drug and drug-related molecules in a sample. All analytical results depend on the measurement of a physical property of an analyte which varies in a known way with the concentration of the analyte. In the case of HPLC analysis the physical property is the peak detected due to the analyte and the measurement used is either the peak area or the peak height. There are a number of different approaches to quantification for HPLC analysis, the most appropriate approach will depend on the method being used.

Calibration Types

The most common types of calibration used for HPLC pharmaceutical analysis are external standard and internal standard.

External standard

This is the most straightforward calibration type. A standard of known concentration is prepared and analysed alongside the sample by HPLC. The response for the standard is compared to that of the sample, and thus the concentration of the sample can be determined. There are a number of requirements when using this type of calibration:

1. An analytical reference standard of known purity is required. Usually the drug molecule will be used as the standard and should be available in sufficient amounts, a purity determination will have to be carried out. For established products certified reference standards may be purchased.

2. The detector should have a linear response over the concentration range expected. The detectors discussed in Chapter 4 have a large linear range, the concentrations used in the method are selected to suit the detector.

3. The concentration of the standard should be similar to that of the sample. This ensures that the sample response is within the linear range of the method. For samples where the concentration is unknown a suitable dilution may be estimated and then the sample injected. If the response is out of range the dilution can then be modified to suit.

4. The dilutions for preparation of the test solutions and the injection volume should be reproducible. The use of volumetric glassware and modern HPLC injection systems enable high reproducibility and accuracy.

5. The recovery of the samples should be 100%. Drug substance samples and straightforward drug products usually have 100% recovery. However, with complex drug products and biological samples it may be difficult to achieve 100% recovery.

Internal standard

This type of calibration similarly requires the preparation of external standard solutions but in addition a constant concentration of a second compound is added to each sample. The sample concentration is directly proportional to the ratio of the analyte to internal standard. The requirements 1, 2 and 3 for external standard also apply to internal standard. In addition the internal standard component should:

1. Be chromatographically resolved from the analyte but elute closely.

2. Produce a detector response similar to that produced by the analyte for a given weight.

3. Be stable in the test solutions.

Internal standard calibration is used in other chromatographic techniques to correct for differences in the injection volume. For the modern injection systems used with HPLC this correction is not required and in this case the internal standard is used to correct for differences in the extraction procedure of the sample. Therefore it is often used for difficult sample matrices such as biological samples and complex formulations such as ointments and creams. The internal standard is added prior to extraction.

Single point calibration

During the validation of the HPLC analytical method the linearity of the method will be established. A series of standard concentrations covering an appropriate range will be analysed and the results subjected to linear regression analysis. This results in the generation of the equation of the line,

$$y = mx + c$$

where y is the detector response and x is the concentration and 'm' and 'c' are constants. Figure 30 shows an example of the type of plot which is generated during method validation. The '×' markers indicate the response obtained for a series of standards injected and the best straight line is drawn through them using regression analysis. Due to small differences in the system and the detector response this calibration cannot be used for all subsequent analyses. The standard needs to be injected alongside the samples each time the analysis is performed.

Figure 30 Example of linearity determination

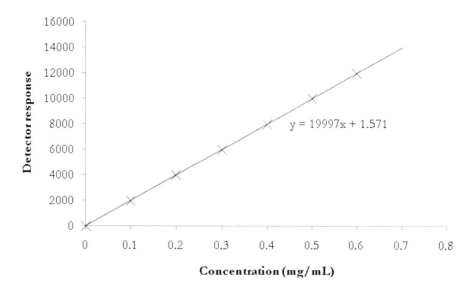

If the value of 'c' in the equation of the line is zero then,

$$y = mx$$

and the line (calibration curve) will pass through the origin. This means that a single standard concentration may be used to determine the value of 'm' and therefore calculate the concentration in a sample. This is very important because it is common practice to use a single calibration standard when analysing pharmaceutical samples by HPLC. In the example in Figure 30 the value of 'c' can be shown to be statistically insignificant and thus equal to zero.

For samples of drug substances and straightforward formulations 'c' is usually equal to zero and single point calibration is normally used. Interferences in the sample matrix can result in a non-zero value for 'c' and in these circumstances the calibration curve will need to be calculated using multiple standard concentrations each time the analysis is performed. Single point calibration is routinely used for external and internal calibration methods.

There are a number of options about which standard injections in a chromatographic run should be used for the calibration of a particular sample. The simplest approach is to use all the standards injected in the analysis to create a calibration curve and quantify all samples against this. Other approaches are to use the standards immediately before the samples for calibration of those samples or to use a bracketing approach where the standards immediately before and after the sample are used. Often local procedures will define which approach to use.

Multi-level calibration

Multi-level calibrations consist of analysing multiple standard concentrations each time the analysis is performed. A calibration curve is generated and then used for

---

quantification of the analyte in the samples. The number of standard levels to be used depends upon the method being used, e.g. if a method requires a non-linear calibration curve then between 6 to 8 levels may be necessary but for a method with a linear calibration curve, 3 may be sufficient. This type of calibration is typically required for **bioanalysis**.

Area Percent

Quantitative results may be obtained by comparing the area obtained for all the peaks in a single sample chromatogram. The results may be expressed in a number of ways:

1. All the peaks in the chromatogram are integrated and each peak is expressed as a percentage of the total area.

2. All the peaks in the chromatogram are integrated except those due to the solvent and each peak is expressed as a percentage of the total area.

3. All the peaks in the chromatogram are integrated except those due to the solvent and each of the minor peaks are expressed as a percentage of the major peak. This is a type of single point calibration. (Assumes that the sample is composed of one major component and other minor components.)

This is a very convenient way to obtain quantitative information for a sample without having to use a reference standard. It is commonly referred to as 'area percent' determination and is used extensively in the early stages of drug discovery and development. However there are limitations to this approach:

It assumes that there is a linear relationship between the amount of each component and the response obtained over the range of the component present in the lowest amount to the component present in the highest amount (usually a drug or intermediate). If a universal detector is used then the peaks will relate to the amount of the component present, however no robust universal detector is currently available for HPLC and the most common detection technique is UV. The magnitude of the peak observed using UV detection is not only related to the amount of the component present but also to the molar absorptivity for that molecule. Even for structurally similar molecules this value can vary considerably and result in under or over estimation of components in a sample.

Components in the sample which do not have a detector response will not be included in the calculation. This includes organic species which do not respond to the detection method (e.g. in the case of UV detection, compounds which do not contain a chromophores), residual solvents in the sample (if applicable) and also inorganic species.

Types of Methods

The most common types of methods used in pharmaceutical analysis are assay and determination of impurities. Each of these is discussed below in relation to the approaches which are used for quantification. Other types of HPLC methods include: content uniformity, analysis of dissolution samples, analysis of cleaning samples and analysis of environmental samples. External standard calibration is the preferred technique for these methods.

Assay

An assay of a drug related sample is simply an analysis to determine how much of the material is present. In the case of a drug substance this is usually expressed as the amount of drug present as a percentage of the total amount of material, % weight/weight (%w/w). This also applies to other single component samples such as raw materials, intermediates and excipients. In the case of a formulated product the assay result is usually calculated as the amount present in a single dosage unit or a percentage of the amount that should be present (i.e. the % label claim).

Usually external standard calibration is used unless a complex sample matrix makes the use of internal standard calibration necessary. Since an assay often involves only one peak the concentration of the standard and sample solutions can be selected so as to obtain the most suitable response for the detector. Ideally the concentration should be the same for standard and sample.

Determination of impurities

The quantification of impurities in a sample may be achieved using a number of different approaches based on external calibration method and also the area percent method. A typical sample to be analysed for impurities will contain the drug as the major component and the impurities as minor components. The level of the impurities will depend on the nature of the sample. Just a few examples of the different types of samples that may be encountered are given:

1. In early stages of development when the synthetic route is being developed the level of synthetic impurities in a sample of drug substance (or an intermediate) may be relatively high. Over the course of optimising the synthetic process these will be reduced.

2. Some drugs are inherently unstable and will degrade quickly to form degradation impurities, usually referred to as 'degradation products'. The levels of these impurities in analytical samples may be relatively high.

3. During development the drug will be subjected to stress conditions of light, heat, humidity, acid/base hydrolysis and oxidation to investigate degradation pathways. These samples may contain high levels of impurities. Also long term stability studies result in the formation of degradation impurities which need to be assessed.

4. For complex formulated products the impurities present in the sample may be due to degradation of the excipients as well as the drug.

5. The detection of very low levels of impurities may be required for impurities that are extremely toxic and thus a trace analysis is required.

The levels of impurities which are permitted in drugs are defined in the ICH guidelines for impurities[1-3]. The amounts permitted relate to the amount of the drug being administered but the lower level at which an impurity needs to be able to be quantified is 0.03 % with respect to the drug present, either in the sample of drug substance or in the formulated product. This value is commonly used as the limit of quantification for HPLC methods in pharmaceutical analysis. During HPLC method development the concentrations at which the test solutions are prepared are selected appropriately. Thus small peaks in the chromatogram below this value may be disregarded.

If there is a potential for very potent toxic impurities (e.g. gentotoxic impurities) then the permissible limits for these need to be evaluated on a case by case basis. The analysis for these is likely to require extremely low detection limits since they will need to be detected at trace levels. HPLC analytical methods for these types of impurities are likely to be designed just to evaluate the impurity concerned due to the difficulties associated with the detection of trace levels of analytes.

Some common approaches for quantifying the impurities in a sample are listed below:

Area percent
The simplest approach is to use the area percent method. The areas of the peaks in the chromatogram which are not due to solvent are expressed as a percentage of either the total area or the area of the peak due to the drug.

Drug substance external standard
An external standard is prepared using an analytical reference standard of the drug substance. This is used to create a calibration against which all the impurities are quantified.

Impurity external standard
An external standard is prepared using an analytical reference standard of the impurity of interest. The calibration is used to quantify the impurity of interest only.

Dilution of the sample
The sample test solution is diluted by a factor of 200 to result in a solution containing the drug at a level of 0.5 % (other dilutions can be used). This solution is then used as an external standard against which all the impurities are quantified. The advantage of this technique is that the impurities are expressed with respect to the actual amount of drug substance in the sample and the preparation of the standard is very simple requiring only one dilution.

Combined methods

When using the previous approaches the concentration of the standard is selected so as to be similar to the concentration of the impurity being tested. This ensures that the calibration is optimised for the amount of impurity being tested. In HPLC analysis of pharmaceuticals it is very common to use **combined methods**. These are used to determine the assay and the impurities at the same time using the same standard for both. An external standard or an area percent approach may be used. It is the convenience of this approach which makes it so popular in busy laboratories.

It is not ideal to quantify impurities which may be as low as 0.03 % using a standard at 100 % and therefore the method validation needs to prove that the calibration is feasible. The system suitability test for this type of method may include a dilution of the calibration standard at ~ 0.5 % and a recovery determination will be performed on this solution to show that the calibration can quantify correctly at the levels expected for impurities (i.e. the amount in the solution will be quantified using the calibration and expressed as a percentage of the known amount).

Relative response factor

The disadvantage of using any calibration technique where a component is quantified using a calibration obtained from a different compound is that the detector response for both compounds needs to be the same to obtain accurate results. As discussed previously this is not always the case when using UV detection. The **relative response factor** corrects for the differing detector response. This may be determined during the method development and method validation process if the identity of the impurity in question is known and a sample of the impurity is available. The factor will then be included in the HPLC analytical method. This cannot be performed if the identity of the impurity is not known.

Calculations

The basic calculation used in quantitative HPLC analysis is:

$$C_{sam} = \frac{A_{sam}}{A_{std}} \times C_{std}$$

where:

C_{sam} = Concentration of the analyte in the sample solution for injection

A_{sam} = Peak area due to the analyte in the sample chromatogram

A_{std} = Peak area due to the analyte in the standard chromatogram

C_{std} = Concentration of the analyte in the standard solution for injection

---

It is possible to use the peak area or the peak height for quantitative analysis but peak area is more commonly used. Both are easily measured using an integration system such as a CDS. Sometimes validation studies are performed using both peak area and peak height and the one which gives the best results is selected for the method.

The other calculations in a HPLC analytical method will be to express the result in the correct format and to convert into the required units, e.g. in an impurities analysis of a drug substance the relative response factor for a particular impurity may have to be applied, also the result may need to be expressed with respect to the amount of drug present, in the units of % w/w. Complicated calculations will usually be included in the HPLC analytical method and after a careful set up in the chromatography data system will routinely be performed by the computer.

Calculations and results are usually expressed using weights but sometimes, particularly in bioanalytical methods, molar concentrations are used where Molar is molecular weight in g/Litre.

Summary

1. The main types of calibration used for pharmaceutical analysis are the external standard method and the internal standard method.

2. Calibration curves may be constructed using single point or multi-level approaches.

3. Area percent provides a convenient quantification method.

4. The most common types of methods requiring quantification are assay and determination of impurities.

References

1. International Conference on Harmonisation (ICH) of Technical Requirements for Registration of Pharmaceuticals for Human Use, Topic Q3A(R2): **Impurities in New Drug Substances**, 2006, www.ich.org.

2. International Conference on Harmonisation (ICH) of Technical Requirements for Registration of Pharmaceuticals for Human Use, Topic Q3B(R2): **Impurities in New Drug Products**, 2006, www.ich.org.

3. International Conference on Harmonisation (ICH) of Technical Requirements for Registration of Pharmaceuticals for Human Use, Topic Q3C(R3): **Impurities: Guidelines for Residual Solvents**, 2002, www.ich.org.

Further reading

D.C. Harris, '**Quantitative Chemical Analysis**', 7th ed., W.H. Freeman, 2006.

E. Katz (ed.), '**Quantitative Analysis using Chromatographic Techniques**', John Wiley & Son, 1987.

Notes

Glossary

A

Accuracy	Expresses the closeness of agreement between the true value and the value found.
Active Pharmaceutical Ingredient (API)	The 'active' or the 'active pharmaceutical ingredient' is the substance in a drug preparation that is pharmaceutically active.
Adsorbent	Packing used in adsorption chromatography.
Adsorption	The process of interaction between the solute and the surface of an adsorbent, e.g. silica.
Analyte	The compound of interest to be analysed by injection onto and elution from a HPLC column.
Assay	An analytical method to analyse or quantify a substance in a sample.

B

Back pressure	The pressure experienced when mobile phase is pumped through the column.
Band broadening	The process of increasing width of the chromatographic band as it moves down the column. Sometimes called band dispersion or band spreading.
Baseline	The region in a chromatogram where no peaks are present and the detector response is due only to the mobile phase.
Bioanalysis	The chemical analysis of biological samples, e.g. plasma, urine etc.
Bioassay	A biological test, measurement or analysis to determine whether compounds have the desired effect either in a living organism, outside an organism, or in an artificial environment.
Biopharmaceutical	A drug produced by biotechnology.
Biotechnology	The application of scientific and engineering principles to the processing of materials by biological agents.

Bonded phase	The stationary phase held in place on the column by chemical bonds with the column packing.
Buffer	A solution that maintains constant pH by resisting changes in pH from dilution or addition of small amounts of acids and bases.

C

Chiral	A compound that is asymmetric, it is not superimposable on its mirror image.
Chromatogram	A plot of detector signal output or sample concentration versus time or elution volume during the chromatographic process.
Chromatography	A technique for separating mixtures based on differential migration of components carried by a mobile phase through a stationary phase.
Chromatography data system	A software package which is designed to enable acquisition and processing of chromatography data.
Chromophore	A part of a molecule that selectively absorbs light at particular wavelengths.
Cleaning	Applied to pharmaceuticals, this is the process of cleaning equipment used in the manufacture of drug substance and formulated products so that the amount of residual drug is below an acceptable level.
Combinatorial synthesis	A method of synthesising large quantities of compounds in small scale using automated or semi-automated processes. Normally carried out as solid phase synthesis.
Combined methods	HPLC analytical methods where the assay and determination of impurities are performed together using the same calibration standard.
Column	A tube (typically made of stainless steel or PEEK) which contains the stationary phase where the separation of the sample takes place.
Critical pair	The pair of components in a mixture which elute closest together and thus if the resolution of these two is acceptable then all other components will be adequately resolved.

D

Degradation products   Compounds which are formed due to the degradation of the drug molecule.

Denaturing   A protein is denatured when its tertiary structure is changed, the tertiary structure can be thought of as the overall, unique, three dimensional folding of a protein.

Differential migration   Refers to the different rates of travel through a column of different components in a mixture.

Diluent   A solvent used to dissolve a substance.

Dipole   A dipole is a directional property and can be represented by an arrow between an electron rich part of a molecule and an electron deficient part of a molecule.

Discovery   The research process that identifies molecules with desired biological effects, and thus have promise as new therapeutic drugs in humans

Dissolution testing   A standardised method for measuring the rate of drug release from a dosage form.

Drug delivery   The method by which the drug molecule is delivered to the drug target in the body, usually involves the use of a formulated product.

Drug product   The drug molecule incorporated into a preparation which can be administered to patients, also known as the formulated or medicinal product.

Drug substance   The isolated drug molecule.

Drug target   Areas in the cells of the body where drugs attach and the result is a therapeutic effect.

Dwell volume   The volume between the point of mixing of solvents (usually in the mixing chamber or at the proportioning valves in the HPLC system) and the head of an LC column.

E

Efficacy   The effectiveness or ability of a drug to control or cure an illness.

Efficiency   A measure of peak band spreading in a HPLC system. The higher the value for efficiency, the more peaks that can be fully separated in a given time.

Elution	The process of passing mobile phase through the column to transport solutes down a column.
Enantiomer	The mirror image forms of an asymmetric molecule having one asymmetric centre.
Endcapping	A column is said to be endcapped when a small silylating agent, such as trimethylchlorosilane, is used to bond residual silanol groups on a packing surface.
European Agency for the Evaluation of Medicinal Products (EMEA)	The European regulatory authority for testing and approval of drugs.
Excipients	The ingredients which are present in a drug product other than the drug substance.
Extra column effects	The total band broadening effects of all parts of the chromatographic system outside of the column itself.

F

Food and Drug Administration (FDA)	The drugs regulatory authority in the US.
Formulated product	The drug molecule incorporated into a preparation which can be administered to patients, also known as the drug or medicinal product.
Formulation	The details of the excipients and drug substance which are combined to make a drug product.

G

Good Laboratory Practice	Scientific codes of practice that apply to a pharmaceutical company's research laboratories and which are monitored by regulatory authorities.
Good Manufacturing Practice	Scientific codes of practice that apply to a pharmaceutical company's production plants and which are monitored by regulatory authorities.
Gradient elution	A process to change solvent strength as a function of time (normally solvent strength increases) thereby eluting progressively more highly retained analytes.

H

High Performance Liquid Chromatography	The modern, fully instrumental form of liquid-phase chromatography technique that uses small particles and high pressures. Sometimes called 'high pressure' LC.
High throughput screening	An automated method of carrying out a large number of in vitro assays on small scale.
HPLC analytical method	Provides the necessary information to enable a HPLC analysis, may include details of the chromatographic systems and the preparation of the test solutions.
Hydrogen bonding	A non-covalent bond that takes place between an electron-deficient hydrogen and an electron rich atom, particularly oxygen and nitrogen.
Hydrophilic	Refers to compounds that are polar and water soluble. Literally means water loving.
Hydrophobic	Refers to compounds that are non-polar and water insoluble. Literally means water hating.

I

Impurities	Undesirable components in samples of drug substance or drug product.
In vitro	Testing procedures carried out on isolated macromolecules, whole cells or tissue samples.
In vivo	Studies carried out on animals or humans.
Ion-exchange	The exchange of ions of the same charge between a solution and a solid in contact with it
Isocratic elution	The composition of the mobile phase is held constant throughout the analysis.

L

Laminar flow	In a cylindrical tube, fluid streams in the centre flow faster than those at the tube wall, which results in a radially parabolic distribution in axial fluid velocity.
Lead compound	A compound showing a desired pharmacological property which can be used to initiate a medicinal chemistry project.
Liquid Chromatography – Mass Spectrometry	The combination of the analytical techniques, HPLC and mass spectrometry.

Limit of detection	The lowest amount of analyte in a sample which can be detected but not necessarily quantified as an exact value.
Limit of quantification	The lowest amount of analyte in a sample which can be quantitatively determined with suitable precision and accuracy.
Linearity	The ability of an analytical method (within a given range) to obtain test results which are directly proportional to the concentration (amount) of analyte in the sample.
Linear velocity	The velocity of the mobile phase moving through the column.

M

Macromolecule	A molecule of high molecular weight such as a protein, carbohydrate, lipid or nucleic acid.
Mass transfer	The process of solute movement into and out of the stationary phase or mobile phase.
Matrix	The packing in a HPLC column onto which a stationary phase is bonded.
Metabolism	The reactions undergone by a drug when it is in the body. Most metabolic reactions are catalysed by enzymes, especially in the liver.
Metabolite	Any substance produced by metabolism or by a metabolic process.
Method development	The process of selecting a suitable stationary phase and mobile phase system which can reproducibly separate components of interest.
Method validation	A process of testing a HPLC analytical method to show that it performs to the desired limits of precision and accuracy in retention, resolution, and quantification of the sample components of interest.
Mobile phase	The solvent that moves the solute through the column.
Monolith	A HPLC column which is packed using a single piece of polymeric silica gel rather than small particles.

N

New Chemical Entity (NCE)	A novel molecular structure for a drug.
New Molecular Entity (NME)	A novel molecular structure for a drug of biological origin.
Non-polar	Used to describe compounds that do not have a dipole moment. They possess hydrophobic (water repelling) characteristics and are not easily dissolved in water.
Normal phase	Used to describe liquid chromatography where the stationary phase is polar and the mobile phase is non-polar.

P

Packing	The adsorbent, gel, or solid support used in an HPLC column.
Partition	In partition chromatography the analyte partitions in equilibrium between the liquid mobile phase and a covalently bonded organic phase.
Peak	The visual representation on the chromatogram based on the detector's electrical response due to the presence of a sample component inside the flow cell.
Peptide mapping	The characteristic pattern of fragments formed by the separation of a mixture of peptides resulting from hydrolysis of a protein or peptide.
Pharmacodynamics	Mechanisms by which drugs affect their target sites in the body to produce their desired therapeutic effects and their adverse side effects.
Pharmacokinetics	The study of the metabolism and action of drugs with particular emphasis on the time required for absorption, duration of action, distribution in the body and method excretion.
Pharmacology	The study of the properties of drugs and their effects on living organisms.
Pharmacopoeia	A book containing directions for the identification of samples and the preparation of compound medicines, and published by the authority of a government or a medical or pharmaceutical society.

Placebo	A preparation that contains no active drug, but looks and tastes similar to the preparation of the actual drug.
Planar	Planar chromatography is performed in two dimensions where the mobile phase travels over a planar surface of stationary phase to achieve the separation.
Polar	Used to describe compounds that have a dipole moment because they consist of molecules that have negative and positive poles. They possess hydrophilic (water attracting) characteristics and are easily dissolved in water.
Polarity	The property of molecules of having an uneven distribution of electrons, so that one part has a positive charge and the other a negative charge.
Potency	The amount of drug required to achieve a defined biological effect.
Precision	Expresses the closeness of agreement (degree of scatter) between a series of measurements obtained from multiple sampling of the same homogeneous sample under the prescribed conditions.
Preparative HPLC	An isolation and purification tool whereby the components in a mixture are separated on a HPLC column and then collected.
Protease	Enzymes which hydrolyse peptide bonds.
Protein	A macromolecule made up of amino acid monomers. Includes enzymes, receptors, carrier proteins, hormones and structural proteins.
Proteomics	The study of the structure and function of proteins, including the way they work and interact with each other inside cells.

Q

Qualitative	Analysis which merely determines the components in a sample without any regard to the quantity of each ingredient; contrasted with quantitative analysis
Quantitative	Analysis involving the measurement of quantity of the component in a sample.
Quantification	The act of expressing the quantity of something.

R

Relative response factor	A factor used to correct for the different detector response experienced for different analytes using some detection techniques (e.g. UV).
Relative retention time	The retention time of a component in a mixture expressed relative to another component (usually the drug molecule).
Repeatability	Expresses the precision under the same operating conditions over a short interval of time.
Resolution	Ability of a column to separate chromatographic peaks.
Retention time	The time between injection and the appearance of the peak maximum.
Reversed phase	Used to describe liquid chromatography where the stationary phase is non-polar and the mobile phase is polar.
Run time	The time during which data is collected for a single injection and is displayed using a chromatogram.

S

Selectivity	A thermodynamic factor that is a measure of relative retention of two substances. It is fixed by a certain stationary phase and mobile phase composition and therefore can by altered by adjusting these.
Silanol	The Si–OH group found on the surface of silica gel.
Silica	A chemical compound comprised of silicon and two oxygen atoms (SiO_2).
Siloxane	The Si-O-Si bond found in silica or for attachment of bonded phases.
Size-exclusion	A non-interactive chromatographic technique which separates solutes according to their molecular size in solution.
Sonication	Passing high frequency sound waves through a sample, a.k.a. ultrasonication. Used to remove air from mobile phases and aid dissolution of samples.
Stationary phase	The chromatographically retentive immobile phase involved in the chromatographic process.
Stereochemistry	The study of the spatial arrangements of atoms in molecules and complexes.

Structure activity relationship	Studies carried out to determine those atoms or functional groups which are important to a drug's activity.
Supercritical fluid	A liquid or gas under extreme pressure.
System Suitability	A test designed to measure the overall performance of a HPLC system to ensure that it is operating as expected. It is carried out each time an analysis is performed.

T

Tailing	A peak in a chromatogram which experiences tailing will have an extended trailing edge.
Thin layer chromatography	Chromatography where the stationary phase consists of a thin layer on a planar surface and the liquid mobile phase travels over it.

U

Ultra high pressure liquid chromatography	HPLC where the small particle size used in the columns generates pressures of up to approximately 100 MPa (equivalent to approximately 1000 bar or 14500psi).
UV cutoff	The wavelength above which a solvent, buffer etc. is not UV active.

V

Validation	See method validation.
Void volume	The total volume of mobile phase in the column, the remainder of the column volume is taken up by packing material.

List of Abbreviations

ACN	Acetonitrile
ADME	Absorption, distribution, metabolism and excretion
APCI	Atmospheric pressure chemical ionisation
API	Active Pharmaceutical Ingredient
AUFS	Absorption Units Full Scale
CDS	Chromatography Data System
CN	Cyano
CSP	Chiral Stationary Phase
DAD	Diode Array Detector
DMPK	Drug Metabolism and Pharmacokinetics
ECD	Electrochemical Detector
ELSD	Evaporative Light Scattering Detector
EMEA	European Medicines Agency
EP	European Pharmacopoeia
ESI	Electrospray ionisation
FDA	Food and Drug Administration
GC	Gas Chromatography
GCP	Good Clinical Practice
GFC	Gel Filtration Chromatography
GLP	Good Laboratory Practice
GMP	Good Manufacturing Practice
GPC	Gel Permeation Chromatography

HILIC	Hydrophilic Interaction Chromatography
HPLC	High Performance Liquid Chromatography
HTS	High Throughput Screening
ICH	International Conference on Harmonisation
ID	Internal Diameter
IEC	Ion-exchange Chromatography
IND	Investigational New Drug
IP	Ion Pairing
IPA	Isopropanol (iso propyl alcohol)
JP	Japanese Pharmacopoeia
LC	Liquid Chromatography
LCMS	Liquid Chromatography Mass Spectrometry
LIMS	Laboratory Information Management System
LOD	Limit of Detection
LOQ	Limit of Quantification
MAA	Marketing Authorisation Application
MeCN	Acetonitrile (methyl cyanide)
MeOH	Methanol
MS	Mass Spectrometry
NCE	New Chemical Entity
NDA	New Drug Application
NME	New Molecular Entity
NMR	Nuclear Magnetic Resonance
NP	Normal Phase
OD	Outer diameter
ODS	Octadecylsilane

PDA	Photodiode Array
PEEK	Polyetherether ketone
pH	Parts hydrogen (measure of acidity)
PK	Pharmacokinetics
pKa	Dissociation constant of an acid
ppm	Parts per million
psi	Pounds per square inch
PTFE	Poly(tetrafluoroethylene), Teflon
QA	Quality Assurance
QC	Quality Control
QSAR	Quantitative Structure Activity Relationship
RI	Refractive Index
RP	Reversed Phase
RRT	Relative Retention Time
RT	Retention Time
SAR	Structure Activity Relationship
SEC	Size-exclusion Chromatography
SFC	Supercritical-fluid Chromatography
SOP	Standard Operating Procedure
SPE	Solid Phase Extraction
SST	System Suitability Test
TEA	Triethylamine
TFA	Trifluoroacetic acid
THF	Tetrahydrofuran
TLC	Thin Layer Chromatography
TMS	Trimethyl chlorosilane

UHPLC   Ultra High Pressure Liquid Chromatography

USP   United States Pharmacopeia

UV   Ultraviolet

Index

Lightning Source UK Ltd.
Milton Keynes UK
19 May 2010

154347UK00001B/6/P